Y0-BYF-685

PSYCHOLOGICAL TIME
AND
MENTAL ILLNESS

Psychological TIME and Mental Illness

MATTHEW EDLUND, M.D., M.O.H.

GARDNER PRESS, INC.
NEW YORK LONDON

GARDNER PRESS, INC.
19 UNION SQUARE WEST
NEW YORK, NEW YORK 10003

All Foreign Orders Except Canada and South America to:

Afterhurst Limited
Chancery House
319 City Road
London, N1, England

Library of Congress Cataloging-in-Publication Data

Edlund, Matthew, 1954–
Psychological time and mental illness.

Bibliography: p.
Includes index.
1. Time—Psychological aspects. 2. Biological rhythms. 3. Time perception. 4. Mental illness. I. Title. [DNLM: 1. Biological Clocks. 2. Mental Disorders—psychology. 3. Time Perception. BF 468 E23p]
BF468.E35 1986 153.7′53 85-20577
ISBN 0-89876-122-0

Printed in the United States of America
Design by Sidney Solomon

Contents

Preface

The study of time merits greater emphasis in psychological research. With rare, if spirited exceptions, the psychology of time is rarely discussed, mentioned, or considered in psychological or psychiatric practice. This book has several major purposes: (1.) First, it endeavors to show the richness of what is already known about time. Clinicians generally do not know that time is a subject with a crowded research history of over 100 years.[1] (2.) A second purpose is to understand why clinical psychology and psychiatry rarely deal with time as a subject. Partly this may be blamed on the difficulties of classification, which have prevented psychology and psychiatry from differentiating objective (clock) time from subjective (felt) time. Curiously, subjective time is where temporal disturbances characteristically appear as a result of particular psychiatric syndromes. Subjective time itself is a field that describes a highly complex psychological area where research is almost nonexistent. (3.) The present advance of research in human biological rhythms opens up a different way of viewing organisms, a new focus where biological subjects can be investigated regarding questions of environment and personality, all in relation to the subject of time.

Why is time so rarely studied, particularly in psychiatry? The present distrust of fundamental psychological elements in psychiatry is probably one factor. Subjects such as time and space are so necessary to perception that to most practitioners they are too vague to consider. That such subjects have been studied, sometimes on a large scale, is generally not known to clinical psychologists and psychiatrists.

There is another underlying notion about time among clinicians—that such a fundamental subject obviously cannot have much clinical relevance. This idea is incorrect. Frederick Melges[2] has shown how important time is in psychotherapy, especially with questions of interpersonal communication. Rare psychological attention to the time sense, particularly with regard to subjective time, has caused the neglect of facts of some clinical significance. A circadian view of ourselves—an understanding, for example, that haloperidol's effective dose, at least in animals, will vary seven times depending on what hour of the day it is given—demands attention from all clinicians.[3] We need to look at people in a circadian fashion, understanding the importance of our innate biological clocks, in all areas of medicine and medical treatment. It may change our view of what "healthiness" is.

Perception, cognition, and affect all influence our senses of time. What is known about how humans learn their time senses, and how these senses change with age, will be reviewed, as well as the voluminous literature on time and mental illness, and the contributions of psychoanalytic theory. Finally we will consider where to go with time—how research might progress, how methodology will necessarily be difficult, and how our internal biological clocks ultimately may relate not just to affective disease, but to our perceptions and personality.

Historically the empirical psychological study of time flows from work on the philosophy of time. One philosophical figure is consistently prominent—that of Kant. Kant's categories of mind, with time and space defined as a priori cognitions, have endowed an enormous exegetical literature. In our day the opposing camps generally array on two sides, regarding Kant as the progenitor of German idealism, or seeing him as the originator of scientific explanation in the study of time. In his scientific works, particularly his study of nebulae and the origin of planets and solar systems, Kant was a heroic, revolutionary force in the creation of modern science.[4] His influence on the study of time was also critical. We frequently find Piaget[5] and Freud[6], to name only two, publicly asking themselves what Kant might say of their results.

The true empirical, scientific study of psychological time began after the midpoint of the nineteenth century. It was part of the great efflorescence of German science and remained a part of that intellectual advance until the beginning of the next century. In 1860 Fechner wrote a series of theoretical papers considering the limits of human perception. How well could people sense this world with their God-given eyes, ears, and hands? The first major experiments on the powers of human temporal estimation were performed by Mach himself, and possess the expected positivistic rigor. He showed that human temporal discrimination was most accurate at about half a second, a finding that has more or less held. His prototype electric apparatus, later much improved by Neumann, is the ancestor of those used in present-day psychophysiological experiments, especially those studying internal time estimation.[7]

That human time perception is most accurate in estimating a half second may strike one as an interesting fact, and no more. Its significance in the 1860s was somewhat different. The Darwinian revolution had just been launched by its highly reluctant revolutionary. It was sometimes held at the time that if human beings were truly the immediate creation of God, it should follow that human estimates of time should be pretty much the same as a clock. That human estimates were not so accurate, even of the shortest of humanly useful intervals, was part of a large package of unwelcome news. It told members of the nineteenth century intelligentsia a fact some did not wish to hear—that human beings were greatly constrained by their perceptual apparatus. The idea was certainly a very old one, and we can see it at least as far back as Plato's parable of the man in the cave. In the age of nineteenth century progress and optimism, natural human senses might allow people to see stars and galaxies light years distant, but science saw whole universes people were unequipped to appreciate on their own. Only with appropriate mechanical devices could they be envisaged, and the scope and capacity of those machines continued, and continues to build. Faith in the accuracy of human perceptions fell. We have now come to an age in science, particularly in clinical medicine, where machines and devices are often trusted more than our eyes and ears.

Mach's discoveries, though relatively unstartling to our time, created major interest in German scientific circles. By 1868 Vierodt's "law" had been canonized, declaring that people generally overestimate short intervals of time and relatively underestimate long ones. Other conjectures were made, especially one that individuals asked to estimate time would consider "filled" intervals longer than "empty" ones.

Let us briefly describe this debate. Filled intervals give the subject something to see or do; empty ones presumably give a person nothing to see or do. It is the nature of the study of time that such "laws" and conjectures are still actively debated. Experiments are now done, often in the guise of computer-compatible models of human thinking and memory, to test whether these "laws" are indeed true. Another feature of the psychological study of time is that much of the research obtains results which, when attempts are made to replicate them, are confusing, ambiguous, or openly contradictory. Interestingly, many present-day studies of internal time estimation are very similar, down to their undergraduate test subjects, to tests performed in Germany and France in the 1890s.[8]

Toward the turn of the century, while work continued in Germany, major interest in the psychological study of time began in France. Across the channel the English empirical tradition appears to have resisted the lure of time research, at least for a long while. Following the work of Guyau, Bergson's philosophical speculations had a profound effect on French psychology, particularly the psychological theories of time advanced by Janet.[9] In the 1920s, Henri Pieron, followed by his pupil François, noted that people's internal sense of time changed with their body temperature. Concurrently in the United States, Hudson Hoaglund serendipitously reached similar conclusions, led by the perceptual changes occurring in his feverish wife. Thus began the study of people's estimation of time in relation to their biological clocks.[10] Though Hoaglund's finding that a person's ability to estimate time is a direct function of body temperature has been considerably modified, research started in what became a new field: the study of innate biological rhythms in humans. It also gave impetus to the study of the uncanny human capacity for subjective time, our ability to feel that an hour is a minute, or a minute an hour.

The 1930s brought not only Piaget's pioneering work on how the human child begins to investigate, count, and use objective time, but also the beginning of therapeutically oriented clinical investigation of the time sense in states of disease. German neurology dealt engagingly with the subject. In psychiatry important papers were delivered by Aubrey Lewis in England in 1932[11] and in the United States by Paul Schilder in 1936.[12] It soon became apparent that the time sense was severely distorted in several mental illnesses. For the psychoanalytic school, the 1930s also marked a high point of interest in the study of time. Freud remarked that the unconscious was timeless. This timelessness he tried mightily to explain, but by his own admission, he did not succeed. He nevertheless felt that elucidation of the mystery promised "profound discoveries." Though these discoveries have proved elusive, psychoanalytic interest continued with work in the forties by Bonaparte and Fenichel,[13] through the fifties by Joos Meerloo,[14] and from the sixties on by a host of American writers. At least within psychiatry, psychoanalytic interest in the study of time has been most sustained and, in several ways, most thoughtful.

Psychiatric interest in the study of time was rarely keen but in the forties and early fifties, studies in academic psychology continued, increasing in number in the late fifties and early sixties. Myriads of internal time–estimation studies were conceived and run to test an enormous range of variables. Did middle-class individuals really estimate short intervals of time differently from lower-class individuals? Would elementary school students in a poor, inner-city neighborhood, given the opportunity to steal from a teacher's purse, possess a different concept of time than those with the same opportunity who did not?[15] Would "extroverts" estimate time differently from "introverts"?

Thousands of undergraduates were tested; many thousands of questionnaires were sent, filled out, returned. A variety of prospective tests were created in an attempt to understand individual future time perspectives. Though interest has waned since the early seventies, the study of time continues in many areas of psychology and psychophysiology. The majority of these studies continue to investigate the estimation of short intervals of time, the methodology of which was outlined by Clausen in 1950.[16]

Goldstone, Lhamon, Boardman, and their co-workers began (in the late fifties) their complicated, difficult studies of the estimation of very short time intervals, comparing normals with schizophrenics. While at first it appeared that schizophrenics estimated short intervals very differently from normals, many years of work replicated this finding with only certain, very specific methodologies.[17] Indeed the problem of methodology is very prominent in the study of the human sense of time. In the late seventies, these authors eventually concluded that the study of time had become far more involved and difficult to interpret than they had first suspected.

Since the late sixties, a major figure devoted to the study of time has been Frederick Melges. While devising a large series of new tests, he has gone on to postulate time as a major element in his theories of psychopathology, developing cybernetic models of brain and behavior.[18] The psychological study of time also owes much to the encyclopedic work of Leonard Doob.[19]

Subjective time has been studied very little during any period. Indeed it is difficult to quantitate.[20] Ours is quantitative age, and we often trust numbers more than we do words. Whether most clinicians truly understand the Gaussian and Bayesian statistics with which so many studies and laboratory procedures are reported is another matter. If we consider the subject "clinically," we know that relatively few do.

The history of time research is long and rich. This book will particularly consider the following propositions:

1. There are two main ways of experiencing time—objectively and subjectively. With objective time, we compare time's passage with the movement of clocks or our internalized estimates of time. Subjective time is how time is felt, how much time seems to have gone by. It involves our feelings about past, present, and future, and what they mean.

2. Though subjective time has been little studied, and is difficult to quantitate, it may provide more clinically useful information than the study of objective time.

3. Just about anything involving perception, cognition, and affect,—whether personality, sounds or flashing lights, regular or irregular stimuli, pain or pleasure—affects our senses of both objective and subjective time. Our sense of time is generally highly reliant on language and linguistic capabilities.

4. Our sense of time changes with age, and is probably not "adult" in form before the age of 15, if then. Objective time develops out of subjective time, and both senses change rapidly in childhood, continuing to change in later life. What we know about how we learn our time sense owes a great debt to Piaget.

5. Human beings, like all eukaryotes, are circadian animals, and much of our internal biological apparatus is organized around our biological clocks. Although through our behavior we often deny our circadian nature, we are still influenced by it.

6.Time senses change with culture, but all cultures appear to possess objective and subjective time. The remarkably stable human ability to estimate internal time appears in primitive as well as industrial societies.

7. Time needs to be considered more frequently in mental health in clinical work and research, both in our attempts to understand how the brain works biologically, and in psychotherapy and clinical practice.

We will start by comparing objective time and subjective time—the mental clocks we live by, though we so infrequently notice them.

Tempus vitam regit: time rules life. It certainly did rule my life for a while, and I would like to thank those who helped me to understand it more: Drs. Marvin Stern, Laurence Tancredi, Eugene Lowenkopf, and Erik Gann for their advice and suggestions, and Drs. George Ginsberg and Robert Cancro for making it possible. I remember my days at Bellevue Hospital with great fondness. I would also like to thank Dr. Louis Faillace of the University of Texas, Houston, for his support during the time of this book's completion. Their help is gratefully acknowledged; all faults I acknowledge as my own.

1

Introduction

In the late summer of 1752, tensions increased throughout the British Isles. The common people, enraged by the arbitrary actions of the government, finally turned to open rioting, with disturbances reaching their peak on the day of September 3. It was a unique day in British history, as the beginning of September 3 had simultaneously seen the beginning of September 14, 1752. It was the day that saw the introduction of the Gregorian calendar. Millions of citizens became alarmed, even terrified, at the prospect of instantaneously "losing" 11 days of their lives.[1]

In whatever context it is defined, time has multiple meanings, and this remains true when thinking of psychological time. Most lexicographical definitions of time end in some tautological explanation, the word "time" used to describe itself. Because it is so difficult a word, so difficult a concept, most definitions have described time in spatial terms. The Old English roots of our language are no different; time is found to have its ancestors in words denoting stretching, tide, rhythms, and change.[2] Change is an essential condition in our definitions of time. We are intuitively aware of many different times, conscious of past and future, conscious of time passing, inside and outside of us. Intellectually we are aware of time shifting from culture to culture, from subject to subject, individual to individual, moment to moment. In America and the West, so innate is the concept of time in our behavior and reasoning, that we mostly fail to define it. When we do give it attention, if ever, that attention is scant.

Within psychology and psychiatry, time is not at present a prominent topic. Part of the explanation lies in the breadth of the subject, and perhaps more still resides in its definition. Definitions of time vary with discipline. There is no particular reason why the time considerations of the philosopher or the theologian should be exactly the same as those of the psychologist or psychiatrist, and we must trust that they are different. But to know these differences, we must define them. One of the greatest difficulties in the psychological study of time has been the definition and classification of the subject itself. To speak of the time sense of an individual, whether ill or well, without subtyping, without classification, severely strains the ability to compare results. This difficulty has not prevented many authors from writing about the subject in an undifferentiated fashion. Possibly the main reason that subjective time is so rarely considered in psychology, beyond its difficult-to-quantify character, lies in the fact that it is simply not defined apart from objective time. To be asked how much time has passed on a clock is a very different question than to be asked how much time we feel has passed. We may know that an hour has truly passed, though we have felt that a much longer, or a much shorter, period has gone by. How do we answer? This muddle and indistinction between objective and subjective time is a continuous plague to psychological study of the subject.

Numerous ways of classifying time have been devised, and there is often overlap among them. Because of its clarity and ability to designate significant differences, I prefer the classification of Lehmann, which I will modify slightly.[3] His scheme defines four basic modes of experiencing time and its passing, two each for objective and subjective time. For objective time, there are:

1. External time.
2. Internal time estimation.

For subjective time, there are:

3. Subjective time awareness.
4. Subjective time perspective.

How these modes of experiencing time develop, their potential biological origin, and how they are influenced by environment, activity, culture, character, health, and disease, are the subject of this book. Lehmann's classification implicitly recognizes two major components of our time experiences. The first, an objective component, allows us to experience time in reference to clocks. These clocks may be outside us, as in external time, or within us, in the form of learned cognitions based either on external clocks, on internal biological clocks, or on some combination of both. When asked to estimate time without a clock at hand, we turn to these internalized versions of external clocks, and how we answer tests our capacity for internal time estimation.

The second, subjective time, is wholly different, moving forward, and back, and through time, with an indefinite plasticity. When discussing our subjective sense of the present, of now, and of how long "now" feels, we are speaking of subjective time duration. When speaking of our subjective time sense of past and future, and of how the past, present, and future comingle and achieve their relative importance, we speak of subjective time perspective. Subjective time is a consideration of our human oddness, our singularity, our feeling that time moves within us according to conditions that pay only the faintest attention to clocks. For Virginia Woolf:

> The mind of man works with strangeness upon the body of time. An hour, once it lodges in the queer element of the human spirit, may be stretched to fifty or a hundred times its clock length.[4]

For St. Augustine:

> What then is it that I measure . . . something which remains engraved in my memory. . . . It is in you, Oh my mind, that I measure time . . . it is the ripeness that I measure when I measure time. Thus either that is what time is, or I am not measuring time at all.[5]

For the physicist Sir Arthur Eddington:

> Thus we have immediate experience of the time relation. . . . When I close my eyes and retreat into my inner mind, I feel myself enduring. . . . It is this feeling of time as affecting ourselves and not merely as existing in the relations of external events which is so peculiarly characteristic of it. . . .[6]

This strangeness, the sense of subjective time, is difficult to define, hard to quantify, exacting to measure. It is also a distinctly human characteristic, allowing us experiences outside an absolute, linear, Newtonian world. These four experiential modes of time all deserve individual definition. We shall start with external time.

EXTERNAL TIME

External time is the time of geophysical clocks. It is a time that people in Western culture have learned, preserved, reproduced mechanically, and left to physics to measure. Every individual who wishes to function in our society learns this sense of time from early childhood on. Ultimately external time is physically defined, be it through the second hand of a mechanical watch, or the rate of particle production from a radioactive element.

In industrial societies we so live by clocks that we often forget just how recent these inventions are. Though the Greeks deduced a moderately accurate measurement of external time by measuring the flow of fluids, mechanical clocks that worked efficiently for socially useful purposes only appeared in the thirteenth to fourteenth centuries.[7] The use of watches in daily life by most people has been common for little more than 100 years.

There still exists an enormous population that does not live by or measure external time via mechanical clocks. These people gauge time, through observing the daily changes of the cycles of the earth. For a large part of the world, the far older, cruder clocks of night and day, and the time specified by the position of the sun and moon as the earth rotates on itself and flies past the sun, constitute the principal arbiters of external time.

External time is very significant to industrialized people, and through extension, to the study of their behavior. In one of its forms, external time is fashioned as the time of the natural sciences. Though most of us implicitly consider external time as something approaching the objective time of Newton, a time of absolute space and time, this does not mean that the natural sciences have uniform codes for evaluating and dealing with time. Despite the encroachments of quantum mechanics, Einsteinian spacetime, and thermodynamics on classical Newtonian physics, physics has a very different relationship with time than does biology.[8]

Let us consider some accepted ideas about external time. External time is part of a continuum—it is itself continuous. It is also linear. One second is equal to one second any time in history, except at the extremes of Einsteinian space time. These latter forces almost never appear on earth, and then only in the most extreme, technologically induced conditions, such as those that exist in a cyclotron. For all intents and purposes these conditions are not experienced by people, and do not affect them.

Beyond its continuous, linear character, external time is almost infinitely splittable. We may talk of time lengths lasting seconds or nanoseconds (10^{-9} second), days, weeks, or trillion trillionths of a second. For experimental purposes time measures can be decreased to almost any size we like, and that size, no matter how large or small, is equivalent at any point in time.

Objective time has now as its starting point, and continues on into an indefinite future. As easily as we speak of nanoseconds, we speak of billions of years, and do not realize when we do so that the scientific perspective of time stretching onward into infinity is a rather recent development. For much of Western intellectual history, such ideas were literally unthinkable. From the fifth century to the eighteenth, a well-accepted corollary of Christian theology, derived from Biblical exegesis, was the belief that the world would last only 6000 years. If that long.

For well over a millenium, Christian dogma authoritatively decreed that the world was created *de novo* approximately 4000 years before the birth of Jesus. As the world would only last 6000 years, human civilization would end unequivocally somewhere near the finish of the twentieth century (perhaps there may be something to heed in this view). And this was an optimistic interpretation. Luther for one knew the world to be such an evil place, so degraded, so vile, that he doubted its survival a century beyond him.[9] The riots of the English common people in 1752 in this context may not appear quite so aberrant. To "lose" 11 days may not have appeared immaterial when the world could but continue, at the greatest hope, for a few short centuries.

Such a short temporal course defined all areas of human discourse, and physics and geology were no exception. Boyle labored to keep his understanding of erosion and the forces of the earth to the 6000-year limit, and the theory of evolution was impossible so long as nature itself was so temporally constrained. Here is another place where Kant's influence on the study of time is important. Only when he looked through a telescope at the clouds of Maupertui, and saw there not weak bogs of light, but entire galaxies filled with stars, solar systems, and planets, and knew their distance to be great, did conceptions of infinity return to the natural sciences. Our scientific patrimony of infinite time and infinite space, the scope of our scientific imagination, owes much to Kant.

Let us return to external time. Let us conceive it, by reason of a thought experiment, as a line. It is a wonderful Newtonian kind of line, equal from spot to spot, uniform, continuous, infinitely dense, infinitely splittable. External time, the time defined by mechanical and atomic clocks, does not fully allow itself the notion of discontinuities. Time is always flowing. If we wish to look forward or back, we must use time to do so. And we must use other temporal conceptions. The first of these is duration, the amount of time experienced in the interval between one event and another. The interval is defined by the phenomenon we wish to measure—it can be whatever we like. Then we must consider sequence, the order in which the phenomena come, before or behind themselves or each other. Finally we must reckon with simultaneity, itself a special case of sequence. Simultaneity is when two phenomena occur at the same time.

At the same time how we define that time is highly significant. Simultaneity is itself a measure of the accuracy and capacity of our measuring devices. Consider, if you will, a race. It can have any participants desired, skiers, runners, tortoises, elephants. All the contestants are ready to begin. At some arbitrary signal declaring "now," they all start. When they reach the finish, there may be one winner, or there may be two, or there may be several—we may have a tie. Whether or not a tie exists is determined solely by the capacity of our measuring devices. If two competitors arrive a hundredth of a second apart, this may prove a sufficient difference to declare one the winner and the other runner-up. If they arrive within a millisecond of each other, however, we may well call it a tie. It is all a matter of how you split time, and with whom you split it. Simultaneity on joining a line in a bank may be very different from simultaneity in a physical chemistry experiment. Sequence or number may or may not change, but we will only know time according to the measure of the capabilities of our instruments.

In most psychological studies of time, external time is rarely considered, but is accepted as a given. The human capacity to deduce clock time tends to remain a highly stable function, and usually only varies with severe disorders or overt brain pathology. In certain conditions of organic brain lesions, or relatively severe mental retardation, the ability to tell time while in the presence of clocks may well disappear. But this is rare. In functional psychoses the ability to tell clock time is very infrequently compromised, and then almost always in the presence of severe pathology. Evidence of such pathology, particularly cognitive pathology in chronic schizophrenia, may have been underestimated in the past.[10]

INTERNAL TIME ESTIMATION

Internal time estimation is our ability to estimate objective time without the cues of external clocks. It is primarily an ability to estimate the duration of the present, and involves little concern for successions and sequences in time. In this sense it is similar to subjective time duration, which is the subjective feeling of time that the experience of the present evokes.

The fact that human internal time estimation is somewhat different from clock time has only been studied relatively recently. Mach himself, the great nineteenth century German positivist scientist and philosopher, was the first to demonstrate that clock time and what people estimate as clock time are not always equal.[11] However, despite the apparent lack of instruments within us to estimate time internally (a subject treated in the next chapter), internal time estimation is in fact a rather stable human activity. It is also surprisingly accurate, particularly at very short intervals, and in some individuals at long intervals as well. This ability to estimate time internally is felt by some to lie in the overlearning of clock time that occurs in industrial societies. There is some evidence that this ability is more general. Significantly, we learn to differentiate internal time estimation from measures of subjective time. We learn to estimate how long our acts and experiences take, even if we *feel* them to have taken more or less time than we *know* them, by the clock, to have taken. Judging by past experience, and perhaps aided by our assortment of internal biological clocks, we are often quite capable of estimating how long something took, even if we lack a watch.

Internal time estimation is still, of course, more variable than external time calculation. The ability to estimate time is most often severely compromised in individual cases of organic brain syndromes, in toxic-metabolic confusion, and in some cases of acute psychosis. To some extent it may vary with character and affective illness, but as we show in Chapter 5, to a far lesser degree than with the aforementioned categories. It may vary only little with cultural background. To estimate time internally is primarily a perceptual-cognitive issue, yet one where disorders of affect, particularly as they involve attention, may play some role. People, via hypnosis or more patterned general learning, can be taught correct or incorrect measures of internal time. When they are taught incorrect time measures, they act on them, and their behavior in the short term can change.[12]

Internal time estimation demands that we confront issues of psychophysiology. Unlike scientific, mechanical apparati that can be reengineered or redesigned, humans are living organisms with severe inbuilt limits of time estimation measurement. Our ability to discriminate different phenomena, and to discriminate meaningfully, is constrained by the capacities of our nervous system.

Our capacities to perceive, and particularly to learn, differ with our different senses. In many ways our hearing is our sharpest discriminant sense. Probably our most subtle discriminant capacity lies in our auditory system's ability to note different sounds. Let us take someone with normal hearing. If two sounds are made in quick succession, an interval of only 2 milliseconds is necessary to allow that normal hearer to know that there are indeed two sounds, and not one. To put them in order is a very different matter; to tell which came first and which came second is a more complicated task, and takes at least 20 milliseconds. Our visual system is several degrees less capable, particularly in discriminating two stimuli as different.[13]

Our ability to make sense of the world is based partly on our ability to define information as discontinuous. As we are highly reliant on our visual systems, much effort has been expended in finding out where and when we view stimuli as continuous, and where we make them out as separate. Visually, this separating-out process takes us somewhere between 20 and 50 milliseconds, the important "critical rate." With a limiting value of around 18 per second, the critical rate has an important function in our daily life, tricking us into thinking the action of a film or television program, for instance, is really continuous. What is usually placed on the screen is about 24 separate photographs per second. Though they are very separate on a roll of film, when they are placed on a projector and shown on a screen, we cannot tell the individual photographs apart. To us they are continuous, and in motion, as we have reached our innate neurobiological inability to separate stimuli.

There are many other limits of perception. One of the oldest problems in psychophysiology is that of the minimal duration of perception—how long a period do we have to have between stimuli to know they are really different? With auditory or tactile actions, the minimal perceptible duration is around 15 to 20 milliseconds. The visual system is much slower, with a minimal perceptible interval of about 75 milliseconds. Our comparative visual slowness allows yet another film trick, that of "subliminal" perception. In the 1930s, this allowed unscrupulous film operators to put "drink Coke" or "eat popcorn" signs into the middle of features. By placing stimuli too quickly to be consciously perceived, they engaged in a very conscious attempt to increase sales. Mixing senses changes the requirements of the minimal perceptible duration. In combining hearing and touch, about 45 milliseconds are needed; 60 milliseconds are needed for vision and touch, and about 90 milliseconds for vision and hearing.[14]

What all this points to is how locked into certain time intervals our perceptions and activities really are. If stimuli are too fast, we cannot discriminate between one set and another; if too slow, we are bored and disinterested, and generally move on to something else. Our nervous system declares exacting limits on our ability to act and be acted upon in time. Over the past three centuries, we have produced a myriad of machines and measuring devices whose perceptual capacities are far greater than our own, but the idea that human perception of time is a highly species-specific element is a pretty old idea. As Coleridge wrote, "The delicious melodies of Purcell or Cimarosa might be disjointed stammerings to a hearer whose partition of time should be a thousand times subtler than ours, [just as] the edge of a razor would become a saw to a finer visual sense."[15]

The literature on internal time estimation is the most voluminous among those studies made of the human senses of time. There are three main, general methods for studying internal time estimation in people. These are time estimation reproduction, time estimation production, and verbal time estimation.

Probably the most stable, though not the most useful way for investigators to set time estimation, is through time estimation reproduction. Here an interval of whatever length and conditions is set for the subject, who is then asked to reproduce it. Often the subject does this by counting silently, by tapping, or by some other regular activity. This ability to reproduce an interval is often easier and more accurate than the method of time estimation production. Here the subject is simply given an interval, often verbally, and told to engage in some activity for exactly that interval. Not uncommonly someone is given an interval of 10 to 30 seconds, and asked to tap it out. The third main estimate used is the verbal estimate. Some time interval is presented, and the person is asked, "How long did you think that was?" The subject then responds verbally.

The permutations, combinations, and methods of comparison for studies of internal time estimation are indeed enormous. The number of test conditions that can vary results are also reported to be numerous. The subject of internal time estimation in disease and "health," however defined, is discussed in Chapter 5.

Human estimation is apparently most accurate around an interval of six-tenths to eight-tenths of a second. Very generally, shorter intervals tend to be overestimated, and longer ones underestimated. Such is a conjecture of nineteenth century German science known as "Vierodt's law." Interval counting of this type has been used to test mentally ill patients for decades, and generally shows most variable values in the acute functional illnesses, as opposed to the "chronic" state. Much of the work of Orme,[16] and of Mezey and Cohen,[17] has shown depressives to do slightly better during recovery as opposed to when ill, with capacities very similar to those of normals. For most people the ability to estimate time remains reasonably stable, especially when groups are compared over variable conditions. One of the more memorable investigations, a Canadian study of the 1960s, took a group suffering "derealization and depersonalization," probably among them schizophrenics and people with personality and anxiety disorders, and compared their time estimation with normals under a series of different conditions. Subjects were rotated around and around in Baranyi chairs with iced calorics placed in their ears, put in sensory isolation for three hours, and even deprived of sleep for 50 hours. Remarkable in several ways, the results of the study noted not only how little these conditions changed the group experience of time, with the possible exception of some measures taken after rotation, but also how comparable both groups remained to each other throughout.[18]

Many attempts have been made in trying to understand, with regard to internal time estimation, whether there is a sort of psychophysiologic definition of the "present," an individual sense of "now." This "now" of internal time estimation is rather different from subjective time duration's "now." It has usually been tested by having people switch off a light or turn off a sound as soon as they feel the "present" to be over. How long the "present" is differs considerably with what type of sensation is being tested, but the variance is always large. Upper limits vary from 2 to 12 seconds and lower limits from 10 to 120 milliseconds, depending on whether you are using sound or light. Our senses come equipped not just perceptually, but experientially, with their own considerations of time, their own conceptions of "now," as well as of interval.

SUBJECTIVE TIME DURATION

Subjective time duration, like internal time estimation, is a measure of duration of the ongoing present. Unlike internal time estimation, however, it is not how much time we think has gone by on a clock, but how long a time we *feel* has gone by. Subjective time duration is the feeling of how time is now passing. Strikingly tied to emotion and affect, it is influenced by perception and cognition. It is also hard to quantify. Because of the difficulty in quantifying subjective time awareness, relatively few studies have considered it, though it is an almost universal sense of human experience. Most of us, when we reflect, remember periods where we felt time to pass slowly, quickly, or normally. Many of us may remember experiences where we felt time to be frozen, or even not occurring at all.

The whole subject of subjective time duration is indefinitely bound up not simply with emotion and affect, but with the question of language itself. Subjective time often has been described poetically and aesthetically, yet it is much harder to describe mathematically. This fact should not surprise us. Consider the word "timeless." Who has used this word, and what has that person attempted to convey? Does it describe an experience where the awareness of time moved suddenly forward and backward into an infinite future and past? Was it the abiding present, the *nunc stans* of ancient philosophy? Does it mean that time has simply no meaning for the condition described? Does it mean that someone was simply not aware of time, consciously or unconsciously ignorant of its passage? Is it the expression of a meditative state, or some existential awareness? Or is it just the polite, hopeful usage of aesthetic criticism, as in a "timeless" essay, or "timeless" beauty? To think of subjective time, we immediately resort to words, though the experience of subjective time must also occur at nonverbal levels.

Subjective time duration is particularly interesting in that severe disorders of subjective time sense may be experienced as the result of a variety of organic brain disorders, while external time calculation and internal time estimation remain intact. Some recipients of bilateral thalatomy have been described as sensing the passage of time as terribly slow, as feeling that an hour takes the space of a whole day, yet they still retain the capacity to estimate time passage and calculate its movement. It is probably correct to say that although these people feel time to pass slowly, inexorably slowly, they *know* that it does not.[19] German research reports of the

1930s described a group of patients whose experience of time was terribly fast, yet who could still calculate time well and knew it to be occurring normally—for others. They were identified as possessing right occipital lobe lesions; some described the phenomenon as a sort of "time lapse," as if everything around them had been filmed by a camera in slow motion, and then projected at normal speed.[20] Other work, with rats, has shown the extrapyramidal area to be heavily involved in the temporal patterning of behavior. Efron, working with time-disordered patients secondary to organic impairments, noted that none of them experienced time distortion difficulties if they did not concomitantly experience aphasia.[21] So we return to the importance of language. The involvement of language in the processing of time may help explain the disarray that occurs in trying to localize time capabilities in the brain. We may do best by accepting the viewpoint of Luria[22], that language is so extremely important to our species that almost the whole cortex is at some point involved in creating and using it. Similar considerations may well apply to the processing of time.

That brain-damaged patients experience subjective duration aberrantly, while retaining internal time estimation and external time awareness intact, points to a common activity of many people in industrialized societies. We are frequently, recurrently, and most often simultaneously, experiencing subjective time duration, internal time estimation, and external time calculation, and comparing them with each other, synchronizing them. We perform some task and feel it to have taken a long time. Somewhere we remind ourselves to estimate internally what it has taken up in time. We then look at a watch or a clock and see how much time it "actually" took. We are concurrently and simultaneously fitting subjective time duration, internal time estimation, and external clock calculation into a generally unnoticed, seamless summation of the experience of time. Yet for all our efforts, we still sense and feel time in a subjective framework as very different from what our external clocks tell us.

One of the main theories of what lies at the heart of subjective time duration involves arousal—the sense of immediate attention and general excitation in doing a task. Like many other qualities, arousal is very hard to define, no matter where it is considered in psychology or psychiatry. Some views of subjective time duration suggest that activity level is a main determining factor.[23,24] Other workers describe a common-sense theory that boredom, at least during the "present," increases the time someone feels has passed, while excitement decreases that sense of duration.[25] Different considerations may point to pain and pleasure as main mediating factors, if broadly defined. Whatever the neurological theories, and several have been described, firm understanding of subjective time in neurobiological terms remains far off.

It is not easy to understand the extraordinary elasticity of subjective time duration, nor its indefinite intermingling with language. We must consider the importance of memory when engaging either subjective time sense, subjective time duration, or subjective time perspective. There is something about our memories that juggles subjective time awareness with the changes of the clock, pointing to yet another consideration: that our sense of time, and how it has passed or gone by, is highly significant for the content and meaning of our lives. With subjective time there may be occasions where what is meaningful to us in fact does "take the most time"—if not on a clock, then in our more personal measures of importance.

SUBJECTIVE TIME PERSPECTIVE

Subjective time perspective is the fourth mode of experiencing time. It exposes an unusual human trait, our ability to trace our lives forward into the future and backward into the past. Subjective time perspective may encompass memories of internal time estimation, or images of future subjective time duration. It is also more directly concerned with the logico-arithmetic considerations of a cognitive external clock, laying past and future out in some form of ordered sequence. Most important it is concerned with meaning, with emotion and affect. Subjective time experience tells us how we personally read the future and the past. Subjective time perspective does not follow the dictates of external clocks, though it does compare our inner feelings with the lines of external time. When someone says, "I feel as though it happened yesterday," or remarks on how fresh a memory or anticipation feels to the mind, that person is talking of a subjective time experience. How we consider and measure the parts and totality of our lives, how we place ourselves in society, are characteristics very structured by our sense of subjective time perspective.

Thoroughly difficult to quantify in any measure, and generally approached indirectly, and through parts, most people agree that subjective time perspective markedly changes with age. To a small child, the future is enormous and infinite, a year from now an almost unknowable amount of time. For the elderly a year may represent only a short duration, too quickly over, too quickly forgotten. Numerous attempts have been made, with subjects of varied ages, to get at this sense of perspective. One of the simplest methods whose use may be clinically applicable has individuals take a strip of paper of some specified size with a line on it perhaps 10 centimeters long, and then has them mark off where they feel one week ago to have occurred, last year, the time they entered school, their birth, and sometimes their expected death.

Of course, time is not parcelled out according to the dictates of an external clock. Now and yesterday invariably constitute a highly inconsistent proportion of someone's life. The recent past is generally seen as long, and is generally far out of proportion to the full measure of one's life. The remote past, particularly with the elderly, receives relatively short shrift. With age-controlled populations, some researchers claim that the year just past is arranged by people logarithmically, with that period closest to the present taking up a proportionately greater slice of the line.[26]

An accounting of this sense of time perspective is relatively unusual in medicine, which may be unfortunate. Consider the five-year survival statistics often shown to cancer victims, as part of the informed consent procedure describing alternative forms of therapy. Rarely are year-by-year breakdowns of survival characteristics given to patients. Different treatments often give very different results, as, for example, in lung cancer. One form of treatment may afford 80 percent survival at two years, but virtually no chance of survival at five. Another may offer only 50 percent survival at two years, but 30 percent survival at five years. Different patients may prefer either the first or the second form of treatment, but are often not given the full curves of such survival probabilities.[27] The question, of course, is a very difficult one, without clearly defined answers. People in their 20s may respond very differently from those in their 80s, but we will not know unless we ask. Such questions highlight the fact that our subjective time perspective does not follow a strict, continuous line, and we value our years and our time differently, depending on our age and place in life.

Many other factors besides age affect subjective time perspective. Acute psychosis often leads to relatively dramatic changes, particularly the disorganization of past, present, and future that occurs in some schizophreniform reactions. Other shifts occur depending on personality, character, and the affective components of disease. Indeed, in depression, clinicians talk of the future as "blocked off," and sometimes cognitive approaches are pressed to break through such blockages.

If we return to the convention of a time line, we can see how subjective time perspective is very different from external time calculation and internal time estimation. The line of subjective time is not linear—far from it. Its major characteristic is indefinite plasticity. A single second now may feel far more important than a whole year later on. In near-death experiences, years of the future and past may be felt to have been lived in what is externally calculated as a matter of seconds. Nor do we experience subjective time as particularly continuous, though we usually see it in some form of continuum. Our emotions push us first one way, then another. We do not experience our lives as simply continuous, but conceptualize life as a series of discontinuous breaks, of different "objective" lengths, where something happened to us. And our sense of the importance and priority of past memories and future expectations may not follow any sequence at all. When we think of subjective time perspective, what is most important and most immediate to us may just as well have occurred, historically, 10 minutes or 10 years ago. We may even lose the thread of objective sequence. The importance of events to us subjectively may have little to do with their relative closeness or remoteness in objective time. Sometimes, particularly in psychosis, this priority of subjective time perspective may entirely overwhelm the ability to place events in objective time, and yesterday, tomorrow, or even childhood, may turn and occupy all of the present.

Social distinctions clearly play on considerations of subjective time perspective. Claims have been advanced that social class strongly affects subjective time experience.[28] Similar grounds have been used in trying to describe the phenomenon of delayed gratification. The impact of cultural differences cannot be denied. We describe this further in Chapter 4, where different cultural conceptions of time are discussed as they impact on subjective and objective conceptions of time, particularly those of so-called "primitive" societies.

These then are the four basic modes of experiencing time that are to be described. External time calculation and internal time estimation combine to create our sense of objective time, or time that is continous, and linear. Subjective time duration and subjective time perspective combine in what we may call subjective time, that human process whereby time loses its linear character and shifts forward or backward, increases or decreases in size, in relation to perception and cognition, but most notably affect, emotion, and meaning.

But first we will look at how humans biologically bind time—through their internal clocks. Our inner clocks, most often circadian, are often postulated as basic to our sense of internal time estimation. They are also much more. Not only do they allow us to adapt to future changes in the environment, but they set the changes that move us, biochemically and physiologically, minute by minute throughout the day. The evolutionary roots of our internal, objective clocks go far in explaining their uses for human beings, and their potential focus and significance in illness and health. They also help point to some aspects of human uniqueness that allow our species to use time as no other can.

2

Man and His Biological Clocks

Almost all the organisms on this planet are circadian organisms. At some point in their evolution, all of them have adapted to the cycle of light and dark as a part of their internal environment. Human beings are no exception. To a degree that is important and often overlooked, humans are circadian animals.[1] Our internal processes follow the same circadian, 24-hour patterns as do most organisms, in an area of biology that has been subjected to systematic research for only a little over 30 years. Our rhythms are stable in environments that lack the infinite number of time cues we receive in our social life. Human beings, unlike almost the entire animal kingdom, are not only circadian animals, but animals who through clocks and culture have in many ways transcended the circadian inheritance of evolution. Both our inner clocks and our ability to ignore them are significant for our objective and subjective time senses. Through our circadian rhythms, and the cultural conditions we use to enforce them or overrule them, we have created ourselves as animals that bind time.[2] This chapter is presented in two parts. The first is devoted to people as circadian animals, particularly as to what these rhythms are, where they come from, and how they affect behavior and disease. The second part examines how we are more than circadian animals, how these relatively recent changes have evolved, and how our senses of time have been importantly implicated in the changes.

WHY DO WE HAVE RHYTHMS?

A variety of rhythms exist throughout life, most of which set the organism in line with the geophysical changes of the outer world. Though interest in biological rhythms can be traced back 2500 years to the Greek poet Archilocus,[3] it was only in this century that it was fully realized how many of these rhythms are fully internal, operating autonomously of the external environment. Over the past 50 years, experiments have been done that, by isolating organisms from their familiar environments and placing them in ones without cues, have shown the regularity and comparative indefatigability of many of these internal rhythms. In general rhythms are classified into three categories, in reference to the 24-hour day. Rhythms of less than 20 hours are considered infradian. Rhythms greater than 28 hours in duration, including rhythms that are circannual, or continuing over many years (like the seven-year cycle of the cicada or the 28-year cycle of the western budworm), are termed ultradian. Most research has concentrated on those rhythms of 20 to 28 hours that entrain with geophysical influences to the 24-hour cycle of the day—circadian rhythms.[4]

Starting with the simplest eukaryotic organisms, circadian rhythms primed by internal biological clocks are ubiquitous throughout life. Their inborn, genetic mechanisms are often remarkably precise. Tested away from their regular environment, some cycle lengths in rodents show a standard deviation of about three minutes.[5] Genetic changes lead to changes in circadian receptors, as can be seen in fruit flies or yeast.

It is not intuitively obvious as to why almost everything that can be measured in living things follows a circadian rhythm. Certainly one of the most significant environmental facts about most organisms is the revolution of our planet in its travels around the sun, and the daily 24-hour cycle of light and dark. Unsurprisingly light is often the most prominent *zeitgeber*, or entrainer, of biological rhythms. Yet this does not explain why examples from cortisol to temperature to mitotic activity persist in 24-hour cycles, regardless of the effects of light or dark. Most rhythms, although they may respond to changes of light in the environment (and very many do not), still perform with remarkable periodicity when these entraining factors are gone.

We must look to the evolutionary adaptation of circadian rhythms in general: Why are such patterns fixed in the genetic makeup of organisms? It appears that these timing systems improve an animal's capacity to persist, by allowing it to adapt internally to forthcoming environmental changes. Rather than have a bird simply face the shock of the seasons, or have zooplankton adjust externally to night and day and the changes of temperature and light, the organism adapts within. By following its own clock, the 24-hour function of the day is acclimated to in advance. This still does not explain why DNA replication, given constant environmental conditions, still occurs at a particular time of day.[6] Nor does it explain why generation times in eukaryotic systems, like us, generally occur in the time scale of a day, about two orders of magnitude slower than prokaryotes.[7] A possible explanation, that the circadian penetration of all our functions is an artifact of evolution, is suggested by Colin Pittendrigh:

> It could well be that the tempo of eukaryotic life is not so much a necessity imposed by its additional complexity as a consequence of historical commitment to the day outside as a general *zeitgeber*. Once the environmental cycle becomes 'pacemaker' for even part of the system, selection is likely to adjust other rate constants, bringing the period of the entire cell cycle into the range of daily entrainment. Too many circadian rhythms lack any clear relation to either hazard or opportunity in the external day for us to ignore the pacemaker's role as a purely arbitrary clock serving the temporal organization of eukaryotes.[8]

Not all human rhythms apparently are circadian. In people there is evidence of cycles of 90-minute duration in rapid eye movement (REM) sleep, oral activity, gross motor activity, performance, and renal excretory activity.[9] There is also some evidence for a seven-day periodicity in urinary excretion of 17-ketosteroids.[10] Other evidence suggests rhythms of about 21 days in testosterone excretion. Definite seasonal rhythms in suicide, conception, and mortality can be noted, though they are often obscured by sociocultural influences. The last mentioned factor represents an aspect of humans' continuing disengagement from our biological rhythms—industrialization seems to have led to a "deseasonalization" of mortality rates. Perhaps the most remarkable yearly rhythm is that of suicide. Aschoff has gathered data on suicide for three rather dissimilar countries—Sweden, Germany, and Italy. Not only do the acrophases, or highest amplitudes, of suicide rates occur at the same time in the three countries (highest in May and December), but the data also show the periodicity in all three countries, was the same in the 1870s as in the 1950s.[11] This is in line with well-known data on a similar distribution of depression and manic-depressive illness,[12] and provides some evidence for the hypothesis of manic-depressive illness as a dysfunction of biological rhythms.

Yet most human rhythms are circadian. Blood pressure or heart rate, cortisol secretion or temperature, intraocular pressure of the eye or the ability to perform memory tasks, or even our capacity for internal time estimation, all follow endogenous, internally created rhythms that conform to the 24-hour cycle of day and night.

The clinical significance of our circadian nature is often overlooked. Give a rat a set dose of haloperidol at 4 A.M., and then give the same rat the same dose on another day at 4 P.M. The aggregate effective result of the same dose will now be seven times as great as formerly.[13] A dose of digoxin given at night has twice the effect as when given in the morning. The time of day you take an antihistamine determines whether its half-time in the body will average four hours or 15 hours; a set dose of Escherichia coli endotoxin will kill 80 percent of a group of mice if given in the early evening but will kill only 20 percent if given at night.[14] Mutagenesis with the same dose of the same agent can vary in rate from 6 to 82 percent, depending on the time of day it is given.[15]

In medicine one is generally amenable to looking at laboratory or other phenomena as following Gaussian or other statistical distributions. We generally do not think of the circadian effects on distribution. However, we do consider the temporal effects of intraocular pressure when looking for glaucoma, or most prominently when looking at hormonal levels in any test of pituitary or hypothalamic function, such as the dexamethasone suppression test. Perhaps it is simply another wrinkle in medical biology, adding complexity to an already complicated field. There is something strange to us in the concept of human beings changing daily, hour by hour, in all their biological parameters, as well as in the character of their performance and behavior. Time-of-day effects may well explain why one gets sick tomorrow and not today; why an infectious agent took hold in one person and not in the person's identical twin; why a drug works or does not work; what is normal and what is abnormal. To get a better idea of what these clocks might do, and we learn more about them constantly, we should try to understand where they originate.

WHAT IS THE INTERNAL PACEMAKER?

Where do these clocks come from, and what is their pacemaker? In the early seventies, two separate research groups (Moore and Eichler, Stephen and Aucker) interrupted the optic tracts and thus blinded rodents in an attempt to abolish the entrainment of their circadian rhythms. To almost everyone's surprise, it did not work. Autoradiographic tracing established soon after that a direct retinohypothalamic pathway existed, which bypassed the optic nerves. The end point of the pathway turned out to be somewhat predictable, but still interesting. It was a small group of nuclei ventral and lateral to the third ventricle, just above the optic chiasm—the suprachiasmatic nuclei (SCN). In the rat each nucleus contains approximately 10,000 small neurons.

The SCN are unquestionably linked to many circadian rhythms. Lesioning of the SCN leads to the disruption of various circadian parameters, including drinking, feeding, sleep, temperature, and ovulation.[16] Yet evidence for the rhythmicity of the SCN themselves, though strong, is indirect. The metabolic rate of the SCN, as well as their neural output, change with a clear circadian rhythm, as has been demonstrated in neural isolation experiments. The actual molecular basis for this circadian rhythmicity, presently a hot topic of research, is unclear. Some evidence suggests cyclic AMP activity, while other models report changes in protein synthesis.

Once one moves through the evolutionary scale to primates, a group which includes humans, the SCN do not control everything. If the SCN are lesioned, body temperature and cortisol rhythms, some of the strongest in human beings, are not abolished.[17,18] These rhythms are among those implicated in affective disease. Indeed most models of human circadian rhythms argue for at least two independent oscillators, as do mathematical models.[19] Yet before we dismiss the importance of the SCN to mental illness, we should consider some of its neural projections: to the medial preoptic area and lateral septum of the limbic system, and to the large afferent projection of serotonergic input from the midbrain raphe.[20]

It is one thing to speak of the anatomical basis of biological clocks, but an equally important consideration, from the standpoint of disease, is that of entrainment. What is the input that causes our circadian rhythms to assert themselves? In most animals light and temperature are the dominant agents. In humans something rather different happens, and it may be that for us social cues possess great importance.[21] Blind people follow exact 24-hour rhythms, as do the sighted, and so far the only difference that has shown up systematically is in 24-hour urinary epinephrine.[22]

One can best study endogenous rhythms by taking away all distinguishing environmental and social cues, and observing them in their basal state. When a rhythm is denied the cues of the environment, and denied those factors that usually entrain it to a 24-hour rhythm, it is said to "free run." The study of human biological rhythms under free-running conditions is now about three decades old. Such experiments must be performed in artifical environments, such as caves or widowless, clockless apartments, where light–dark cycles can be artificially ordered, or left to the personal discretion of the subjects. The first major experiments looking at our biological rhythms took as their subject the nonwaking state.

SLEEP

Sleep has probably received the greatest amount of research attention in studies of free-running humans. Some of what has transpired from sleep research, like other areas of chronobiological research, provides evidence for positions opposed to common wisdom, or what we might "logically" expect.

Some of the most interesting research on human rhythms began with experiments in Germany during the fifties. When stripped of all probable time cues, humans showed free-running rhythms that were a little longer than 24 hours. This is not different from experience with most animals, which in free run show periods that are almost never 24 hours, but generally one or two hours more or less than that. Evolutionary advantages have been postulated for natural, free-running rhythms that do not precisely follow a 24-hour pattern.[23] More remarkable is how easily, once entrainments begin—be they light or temperature in animals, or social cues and clocks in humans—24-hour cycles assert themselves.

In their series of subjects, Aschoff and his co-workers found temperature and the sleep–wake cycle to run about 25 hours, with the standard deviation only about a half hour. When people remained in free-run conditions longer, however, odd things began to happen. A negative correlation was discovered between the duration of activity and the duration of sleep, very much contrary to the common-sense of things. Subjects who got less sleep continued to stay awake for longer and longer periods, desiring less and less sleep time. The temperature rhythm did not change, and was hardly entrainable at all. Shifting periods of light and darkness, allowing or not allowing people to use reading lamps, and a variety of other maneuvers had little effect on the temperature rhythm, which still ran for a period set at about 24.8 hours. But the activity rhythm continued to split off from the temperature rhythm, and finally, in many individuals, from the rhythm of sleep. It became possible artificially to entrain individuals to sleep–wake cycles of 30 hours, or even longer.

Thus was discovered the first example of "internal desynchronization" in human beings. We experience it most commonly whenever we take long plane flights. What these early free-running experiments began to show was the uncoupling of the temperature and sleep–wake cycle, or sleep–activity cycle. How frequently they uncoupled, and how much, was strongly related to age. By Aschoff's definitions and according to his data, uncoupling occurred in 22 percent of men aged 17–34, and in 70 percent of those aged 41–71. It seemed to increase in a generally linear fashion with age. The odd part was the negative correlation of sleep with activity.[24]

Generally when people work harder, need more time to accomplish tasks, and stay up later to do them, they sleep longer and later in the morning. This did not happen to people in free run. Not only did people who stayed up longer sleep less, but the general 24-hour period of the day began to lose its meaning. When allowed to function as they wished, some people worked in sleep–activity cycles of up to 50 hours, with perhaps a quarter of it or less spent in sleep.

The main explanation to interpret these findings has come out of experiments performed in the late seventies in this country.[25] These researchers propose that despite its seeming uncoupling from sleep–activity, the temperature rhythm retains great influence on when these activities occur. It was found that REM, REM latency (how long it takes to begin REM sleep), when people wanted to go to bed, and how long they slept were very well correlated with body temperature.

For those subjects who experienced internal desynchronization of body temperature and sleep–activity rhythms, it was found that regardless of the length of their sleep–activity cycle, whether 25 hours or 50, they tended to go to sleep when their temperature cycle hit its low. Even how long they slept correlated extremely well with where their body temperature was when they went to sleep. For a group of five men with internal desynchronization, with their activity cycle split from their sleep cycle, initiating sleep at the nadir of their temperature cycle showed them to sleep about 7.7 hours. The closer in time to the temperature peak they fell asleep, the longer they slept. If they happened to fall asleep at their temperature peak, the average length of sleep was about 14 hours, nearly twice as long as when they went to sleep at the temperature low. How long they were awake had little influence, compared with temperature, on how long they stayed asleep. Even at sleep–wake cycles of 50 hours, people tended to sleep in six- to ten-hour periods, separated by an occasional very long nap.

The importance of temperature on the nature of sleep may help explain a long-known observation—that naps in the morning are characterized by far more dreams and dreamtime than naps in the afternoon. Dreams occur generally in REM sleep, and REM sleep and REM latency (the latter an interesting measure for affective disease research), correlate well with the temperature cycle. The reason you may dream more during a morning nap than during one in the afternoon is that your temperature cycle is nearer its low point in the morning.

An interesting question is just how important internal desynchronization is. How bad is it to have one's internal rhythms out of phase? Most research in circadian rhythms describes a very strong internal coupling mechanism for these rhythms, a coupling independent of strong environmental cues.[26] When monkeys are entrained to light–dark cycles, an eight-degree reduction in outside temperature has no effect on their own body temperature. Things change when animals are placed in conditions without external time cues, and where desynchronization of circadian systems has occurred. Here the same animals may show a decrease in internal temperature of 2°C.[27] Other data, using chronic external desynchronization of light- and dark-phase shifts, show longevity decreases of 5 to 20 percent for insects and mammals.[28]

In humans, activities that tend to produce internal desynchronization are long airplane flights and shift work. With such phenomena the strongly coupled human circadian rhythms uncouple, separating in individually unpredictable fashions, with some rhythms phase-advanced and others phase-delayed. This phenomenon of split human circadian rhythms has been called "reentrainment by partition" by Aschoff. It appears that humans adapt very differently to phase advance, as compared with phase delay.

Much of the research on internal desynchronization has concentrated on shift work. British studies of shift work showed vastly increased numbers of "reentrainment by partition" in phase-advanced shifts, as compared with phase-delayed, shifts. It appears that, at least with the human sleep–wake cycle, phase delay is much easier to adjust to than phase advance.

Over a quarter of the U.S. work force engages in shift work. Though only recently studied in any detail, reports of insomnia and gastrointestinal upset have long been noted among these workers. One of the more interesting "treatment" studies took place when Czeisler and his colleagues went to a Utah potash manufacturing plant. For many years workers had changed shift each week, from the morning shift (8 A.M.–4 P.M.), to the swing shift (4 P.M.–12 midnight), to the graveyard shift (midnight to 8 A.M.). Twenty-nine percent of the workers reported having fallen asleep on the job during the previous three months.[29] Recognizing that experimentally workers adjusted far better to phase delay than to the phase advance engaged in at the plant, the workers were separated into two cohorts. One remained on the swing shifts as of old; the other was placed on a new, experimental shift that moved by phase delay three hours every three weeks. Complaints about work decreased dramatically, self-rated health indices improved, job turnover was less, and productivity increased. Despite its common occurrence, the effects of shift work on marriages, affective illness, insomnia, and self-perceived well-being are at present not known.

On long jet plane flights, "reentrainment by partition" has also been noted, and tends to be much more common on eastbound flights, which are "phase-advanced," as compared with westbound flights, which are "phase-delayed." One of the more unusual reports in this area is an epidemiologic one compiled by two British psychiatric consultants at Heathrow airport. Looking at a series of cases during a six month period, they diagnosed depression more frequently after westbound flights (phase delay) than after eastbound flights. Such a result is directly contrary to what would be expected on the basis of laboratory experience, and merely points out that human behavior is a much more complicated business outside laboratories than inside. No data on sex distribution, age, or previous diagnosis are given by the British researchers. One soon realizes that the westbound "depressives" include a large number of Britons returning from Europe, of whom many are coming back from holidays. For them the escape from daily life may not have been long or replenishing enough. Numerous other conjectures may be made to explain the data, but given the myriad of variables involved, we cannot know. Jet lag is a major problem not merely because rhythms fall out of phase with each other, but because the cortisol rhythm may take one or two weeks to return to its former state.[30]

PERFORMANCE

One prominent researcher has defined performance as "measures or scores of efficiency at various tasks that require the use of cerebral processes in responding to specified sensory information by appropriate motor actions."[31] Perhaps such a severely behavioristic definition is in order, for performance in humans is notoriously problematic to study well, particularly with regard to circadian rhythms. First problems are the twin factors of motivation and fatigue. A series of other confounding difficulties has led to a severe lessening of possible research designs for investigating circadian regulation of performance, and may well have limited the inferences that can be drawn from such studies. The fact that much of the work has been done at the behest of defense departments interested in the night capabilities of their officers has also tended to skew research design. Nonetheless circadian effects on performance probably exist, and are of some interest.

During early studies on humans engaged in free-running rhythms, time-of-day effects were noted for internal time estimation and simple computational ability.[32] Other factors were then looked at, with greater emphasis placed on the possible impact of biological clocks on behavior. Studies done with shift workers have often shown large differences in the performances of simple mental tasks.

Perhaps the most interesting circadian effect on performance in humans occurs with memory. Much of the work regarding this question has been done with shift workers, and a relatively consistent finding is that short-term memory and long-term memory do not remain the same over the day. In fact short-term and long-term memory appear generally to alternate in efficiency.[33] There is a clear falloff in short-term memory efficiency in the afternoon, when long-term memory appears to peak.[34] As night begins and then continues, short-term memory improves, and long-term memory becomes worse for material presented during the night.[35] Though similar results seem to occur in free-running conditions, it is unclear whether these relationships are the result of endogenous biological factors. Given the huge number of social cues, inference to the biological level is difficult.

Such changes in short-term and long-term memory should be of some interest to pedagogy. If long-term memory is really better in the afternoon and early evening, it is perhaps curious that most teaching time occurs in the morning. Colquhoun points out that some of the original reasons for morning classes were secondary to the assumption that memory was better in the morning than in the afternoon. This may well be true of short-term memory, and given an era of multiple-choice tests, may not be seen by some as a disadvantage. To study the subject of school learning properly, while attempting to control for social confounders of biological effects, would probably be impossible, as well as unappreciated by most educational employees and parents.

A variety of interventions have been made to deal with the confounding difficulties of motivation and fatigue. One interesting research maneuver attempted to note whether motivation would change the circadian rhythm of scores. Taking advantage of the fact that his subjects knew each other, one investigator "motivated" them by announcing individual scores before the assembled subjects, in an attempt to induce "healthy" competition. It worked.[36] Motivation did indeed change results, with the effects most marked at those times of day when scores were at their lowest. The effect when people had their best scores was far less notable. Clearly circadian rhythms are here a contributing factor regarding adaptation, and yet perhaps not the major one.

Several attempts have been made to relate biological performance rhythms to personality factors. Such studies have generally used scales devised by Eysenck differentiating "introverts" and "extroverts," while attempting to correlate performance with body temperature rhythms.[37] Little inferential material can be gleaned from these studies. Perhaps a more interesting notion is that shift workers engaged in nighttime work would show conflict with the social *zeitgebers* "entraining" the surrounding population. From this hypothesis it was reasoned that shift workers isolated from other people would show more rapid adaptation of their performance rhythms than if they immediately went back to "normal time." When shift workers were isolated in this manner, performance rhythms did adjust more fully to normal patterns, but still had not returned completely to normal at ten days.[38]

Circadian performance rhythms are probably more amenable to study in animals than in humans. Electroconvulsive therapy given to mice immediately after they had been trained in certain tasks disrupted memory equally at various points in the day. However, ECT given three minutes after training disrupted memory only in mice tested during the dark phase, and not in those tested in the light phase.[39] Other animal evidence also suggests that memory consolidation may proceed at different rates given different times of day.[40] As the question of memory loss after ECT is indeed a controversial topic, it is probably worthwhile to consider time-of-day effects on memory, particularly anterograde memory.

AFFECTIVE DISORDERS

Current since the 1960s, the theory that affective disease is related to disorders of biological rhythms has been given this public declaration:

> There is evidence that patients with manic depressive psychosis may differ in internal desynchronization of several body rhythms, that is, such rhythms may not exhibit the same phase relationship to each other as seen in normal individuals. Such desynchronization appears to be secondary to a shortening of the periodicity of some of the parameters studied.[41]

Mania and depression have been known as seasonal since ancient times. Recent epidemiological research tends to show mania and depression peaking bimodally in the year, during the spring and late fall. These peaks are curiously similar to the peaks of the acrophase for suicide in several European countries, which have remained constant for 100 years. It has also been known for about a century that sleep deprivation in depression actually helps elevate mood. The effect after one night is transitory and occurs in only a minority of patients, but appears consistent over a variety of populations of depressed individuals.[42] Thus the hypothesis of biological dysrhythmia as a cause of affective disease has a long history, whether or not it is correct.

Much of the significant research on biological rhythms and depression has been performed at the National Institute of Mental Health (NIMH) over the last decade by Wehr, Goodwin, Lewy, Wirz-Justice, and others. Their recent theorizing points to two prominent circadian systems in humans. The first, or "strong" system, ostensibly involves temperature, REM sleep, and cortisol rhythms, all of which are seen as deeply interrelated. This strong system is said to often dominate and propel the "weak" sister system of oscillators, composed of the activity–rest and sleep–wake cycles.[43] Though studies of the activity–rest cycle are notoriously difficult (how do you measure "activity"?), Wehr and his co-workers now believe that significant subpopulations of depressives show circadian rhythms of less than 24 hours. They have produced their most impressive results with bipolar patients, where a variety of circadian rhythms, including motor activity and body temperature, are said to appear to be phase-advanced by one to three hours. One of their more interesting studies was to take a group of rapidly cycling bipolar patients, and to deny them sleep for 40 hours. Under this peculiar stress, seven of their ten patients switched from depression to mania. Various elements of desynchronization were noted. As in all the studies mentioned, the numbers are small, and the data suggestive.

It is particularly significant that in common with virtually all other biological phenomena in humans, neurotransmitters, including serotonin, epinephrine, dopamine, endorphins, and many of their metabolites, have circadian rhythms. These circadian rhythms are phase-delayed by tricyclic antidepressants. Similar results appear for the kinetics of tryptophan beta-hydroxylase.[44,45] Since all neurotransmitters and brain amines studied so far appear phase-delayed by antidepressants, it is hypothesized that some mediating factor, perhaps hormonal, is involved in antidepressant action. Beyond this conjecture, what is clear is that biological clock effects may be intimately related to the presumed neuroendocrinological changes that are implicated in depression. What is cause and what is effect will probably not be known for some time.

With regard to circadian rhythms, perhaps the oddest work of the NIMH group is with women who show a peculiar, seasonal form of depression. Their depression tends to begin at the start of winter, and is characterized by hyperphagia and hypersomnia. Some have written and spoken about this experience as if it were some vestigial prelude to hibernation.[46] Understanding the importance of the light–dark pattern of circadian regulation in most animals, it was reasoned that light effects might be etiologic, and their reversal therapeutic, with this group of patients. Thus these women (and a few men) have become inpatient guests of the NIMH, where they have received experimental treatment with bright, full-spectrum light for two hours or so each morning and evening, combined with alternating washout periods without light. Some excellent clinical responses have so far been reported. If indeed it does work, first guesses about how it does have generally turned on melatonin. This hormone of the pineal, which, while independently acting in birds, appears to be under the domination of the SCN in higher mammals, is exquisitely sensitive to light. The full results of this research are as yet not forthcoming. Clearly much remains to be studied regarding affective illness and biological rhythms, though the results so far are very intriguing.

This brings us to the strange case of lithium. There are many hypotheses as to how lithium works, and its multitudinous effects seem to have a bewildering currency throughout the body. One hypothesis of how it might work rests on its ability to change, generally flatten, biological rhythms. Work in humans shows that lithium phase delays calcium, magnesium, and PTH rhythms.[47] Other work shows lithium to have phase-delay effects similar to those of antidepressants.[48] Another notable effect is its ability to damp phase amplitudes.

Given these results, lithium is now proferred a hypothesized role as the preferred drug to treat "cyclic" disorders. There are numerous periodic disorders with psychiatric implications, including cyclic migraine, periodic hypersomnia, and periodic catatonia, as well as manic-depressive illness. Use of lithium in cyclic migraine supposedly shows it to work.[49] Whether lithium indeed will prove to be the panacea for cyclic diseases remains to be seen.

In finishing our discussion of cyclic disease and biological rhythms, we turn to the other end of the spectrum—to diseases that clearly possess desynchronizations of physiological rhythms. Of such diseases with clear desynchronization, the most obvious cases are in the neuroendocrine category. These include Cushing's disease, where cortisol is high and unchanging; acromegaly, with a similar derangement of growth hormone; and Stein-Leventhal syndrome, with a constant rather than cyclic secretion of luteinizing hormone. Some elements of the ill-defined "premenstrual tension syndrome" may also apply. Of note in these diseases is the considerable toll of psychiatric complications.

While biological rhythm desynchronization may be found etiologic in some fashion in affective diseases, it may also turn out to be epiphenomenal. With perhaps greater confidence, it may be seen that internal desynchronization, particularly hormonal desynchronization, may well be expected to have psychiatric fallout. If the foundations of our biological clocks and their mutually interrelated rhythms are disrupted, it should not be surprising that personal and interpersonal dysfunctions are a concomitant result.

TOWARD A MORE CIRCADIAN VIEW

Following the lead of Koch's germ theory, and Bernard's postulates of homeostasis in the nineteenth century, the dominant view in medical biology has emphasized organismal stability and equilibrium. Our present form of the medical model still emphasizes the restitution of the mechanism to normal, returning it to its state of stability and equilibrium deprived of whatever factor or factors caused it to be thrown out of phase. Similarly, evolutionary theory for a long time was concerned primarily with gradualist change, and pinpointing the first motions of incremental movement. A major interest in ecology was less in how systems changed than in their means of systemic preservation. We can now watch a variation on such stability models through the influence of cybernetics, where machine models are created that attempt to mimic human intelligence and similar complicated cognitive and behavioral problems.

Much as the viewpoint in evolutionary theory is altering, it is time to consider some modification in the viewpoint of medical biology. Both theoretical systems must now become more concerned with the swiftness and systematic nature of change. All organisms, including humans, do not just develop over time and age, but change within the day, from moment to moment. Our present medical model, in its more extreme manifestations, contains many similarities to crude cybernetic constructs, and argues that perturbations in the system have to lopped off, much as a surgeon takes out an appendix. Somehow things will then be right again, and equilibrium, like Adam Smith's invisible hand, will reestablish itself. Such a view of illness and disease is too static. If in evolution the end point of interest is persistence, and the means of getting there the acquisition of a gene pool of great variety to enable maximal flexibility and resilience in response to altered environments, we will have to begin to think similarly about individual human beings.

It may not be correct to define health as absence of disease. Health may be better viewed as some level of capacity to respond to a protean variety of changed environments—biological, psychological, and social. Supernormal capacities in one area may or may not be useful in another; they may, as in overly active immune reactions in tuberculosis, be maladaptive. More complex interactive, circadian-informed models may be necessary to modify our present medical model. What may count in the human who becomes ill is an understanding of how, at just that precise moment, some causative agent appeared that a moment earlier or later might have been harmless. We will have to rethink, along circadian lines, the host–organism relationship in infection. Who is infected and who is not, who becomes ill and who does not, may well be determined by interactions involving the cyclic activities of both organisms, as well as those around them.[50]

The effects of internal desynchronization in shift work and jet lag have begun to be noted; the impact of such rhythmic disruption on affective disease remains to be determined. The necessity to think in a circadian fashion will come most powerfully to considerations of pharmacology. Depending on what time of day it is given, an antihistamine will work for four hours, or 15. Depending on when it is given, the effective dose of haloperidol will fluctuate sevenfold. Radiation given at one moment is without effect; at another hour it causes tumor formation. Chemotherapy effects vary by multiples of five or more, depending on when the drug is given.

In our quest throughout this century for more sophisticated quantitative biochemical tools in medicine, we have sometimes overlooked just what these tools mean. Unfortunately, at least at present, many physicians do not understand the statistical structure on which many laboratory tests are based. Now we will need to add a further complication. Normal and abnormal will not flow just from a Gaussian distribution, but from a circadian one as well. We will have to look at laboratory values along a circadian reference sheet, balanced by the understanding that with severe disease phasing of biological rhythms may markedly revise. Our concept of normal and abnormal will change, as we recognize the more complex, interactional nature of health and disease.

TIME AS AN ECOLOGIC NICHE

Like all organisms beyond the prokaryotic level of organization, humans are circadian animals. They are also far more. Different from other animals, humans are at least as much noncircadian as circadian animals, if we look at their social life. In describing the relative inattention paid to biological clocks on the part of biologists until recently, we have a clue to their evolutionary importance in humans. This importance resides most prominently in the lack of importance of biological clocks, in our capacity to overcome them, even if we pay for their dysfunction with mental illness, and for their disruption with insomnia and psychosomatic complaints.

Much has been written about those characteristics that make humans "different" from the other animals. It is a curious, long debate, generally characterized by overt or hidden theological concerns. In the past language and tool making were described as the "archetypal" human activities. Such conceptions have recently suffered defeats, as we have been able to teach chimpanzees and gorillas the rudiments of language, and have watched a variety of apes utilize tools while acquiring food. We now face the fact that human beings are animals whose differences from other animals are differences of degree, and that we must study other animals and their behavior if we are to understand ourselves.

Yet I would like to provide, as have many others, another point of distinction, once again relative, between humans and other animals. It is in our use of time. One of the great adaptations of human beings, an evolutionary adaptation of culture, has been our ability to move beyond our biological clocks, and establish all of the 24-hour day as our biological niche. Moreover it is a niche used and chosen as we will.

Schoner shows, in an elegant essay, that time is rarely used as an ecological niche by most animals, and when it is, it is generally in some stereotyped fashion.[51] Animals can shift from their endogenous rhythms. Animals hunted by people, such as the Nile crocodile, have changed from diurnal to nocturnal habits. Other species, particularly certain types of migratory birds, may use 24-hour periods for flight without consideration of sleep (another finding that sleep and rest may not be as clearly linked as we commonly expect, in line with studies of sleep and body temperature rhythm). Yet no other animal shows the flexibility of humans in the use of time, our capacity to use any part of the diurnal cycle as we wish, to turn day into night and night into day. Our use of all hours of the day as a flexible ecological niche may be unique. It is quite a distance from our early mammalian ancestors, who perhaps used the night to escape the predations of the day-bound dinosaur reptiles, themselves helplessly wedded to the energy of the sun to control their body temperature.

We may ask what sort of evolutionary means were used to accomplish this transformation of time into an ecological niche. Here one should look to more recent, post-Darwinian conceptions of evolution. R. C. Goodwin, a biologist much indebted to his teacher C. H. Waddington, has conceptualized the process of evolution as utilizing a progression of cognition. He regards evolution as obtaining a sort of "creative intelligence," whereby species learn greater competence, as populations, in adapting to the environment. The conceptualization of a species population as a sort of "cognitive group" is a reasonably new idea in biology, and may indirectly provide some useful theoretical implications.[52]

To get some handle on the time-manipulating capacities of people, a rapid approach may be through Paul MacLean's simplified evolutionary model of the human brain. MacLean, who coined the term "limbic system," essentially sees people as walking around with three rather dissimilar, if communicating, functional brains in their heads. The first, the "reptilian brain or R complex," supplies most of our autonomic functions and needs for maintaining physical life. It includes the spinal cord, midbrain, and hypothalamus. The second component, or limbic system, involving part of the temporal cortex, the septum, fornix, and other areas, is concerned with aggression, emotion, and sexual procreation. Through the mediation of the hippocampus, it is intimately involved in memory, and is central to many theoretical explanations of psychotic disease. The limbic system represents the large part of the paleomammalian, or old mammalian, system. The third and final component of our brains is the neocortex, modulating cognitive powers and fine motor movements. Often overlooked in discussions of the limbic system and schizophrenia is the fact that the limbic system is nonverbal. Its appearance precedes that of sites necessary for linguistic communication.[53]

What is the evolutionary function of the neocortex? What does it do? It is perhaps a theoretical tangle to discuss the evolutionary "uses" of any adaptation. To do so may provide fodder to argue a strong reductionist stance, that everything in the body has its "uses," whatever they are. Though such a concept is blatantly teleological, physiology has used it convincingly as an aid to how the body works. The important point in speaking about evolutionary adaptation is to talk about *potential* uses. What needs to be looked at is how, under certain given or hypothesized environmental conditions, a particular function appears or does not appear valuable.

We have tried to show that many physiological and cellular functions do not "logically" fit a circadian model, and yet, in all eukaryotes, they have fallen into a circadian pattern. What probably happened is that opportunistically, like many evolutionary changes, the circadian pattern was adopted as a basic design structure of animal and plant existence, and most functions placed within its context. Similarly we use DNA over and over for our regeneration functions, rather than engaging other means, such as proteins. The point of evolution is persistence. All physiological uses, in evolutionary terms, are potential uses. The appendix, so long described as a vestigial organ, is now found to contain functional immune lymphoid tissue. Thus its "evolutionary use" is upheld (little comfort to the person with a burst appendix). Our physiological capabilities may be adaptive or maladaptive, given a particular set of external and internal environmental circumstances. Still, the fact that teleologic thinking has proved so useful in biology, that once we find a new structure or entity we eagerly set about finding "what it does," rests in the truth that, much of the time, our physiological functions do have adaptive uses.

In both Freudian and Piagetian schemes of human development, objective time is learned, and comes out of, subjective time. Most theoretical treatments consider that animals have at least a conception of subjective time duration.[54] We cannot ask tigers about their sense of the present, whether an hour feels like a minute or a minute feels like an hour. We do know that future planning, at least as regards conscious delayed gratification, is difficult for most animals. Melges states that he decided to work with humans when he realized, after working with apes for four months, that rigorous training caused them to look forward 90 seconds at most.[55] The concept of a time sense for at least subjective time duration is plausible for animals. But time awareness innately involves a capacity for language. Most animals do not have language, and do not appear capable of it.

There appears to be no clear functional "adaptation" for subjective time duration. Perhaps it is merely a vestigial remnant of other evolutionary movements, even the progress of language differentiation itself. One of the more prominent theories of subjective time duration involves arousal. If we are in a fight-or-flight situation, we have a sense of ourselves speeding up, and external clock time, at least relatively, slowing down. Any reader of thrillers or adventure novels knows the lines, "It must have been ten seconds, but it felt like 30 minutes, or just the rest of my life." Some version often appears at what is the (hoped for) emotional plot climax. Considering this arousal theory, it may be that subjective time is useful to us, allowing us to speed up, if only illusorily, our internal cognitive processes in time of danger. By allowing a feeling of prolonged duration for very short periods, we may be better prepared to consider and develop our thoughts.

Yet another "function" of subjective time duration could be with regard to other cognitive functions, particularly creative intellectual functions. In biographical material, and in psychoanalytic experience, periods of creativity are often described as "timeless." People describe themselves as deeply self-absorbed, concentrating so intensely as not to notice the passing changes in the environment surrounding them. For another speculation on its potential evolutionary "use," prolonged subjective senses of time, giving us a sense of "timelessness" and creative "wholeness," may obtain for us a sense of pleasure in intellectual work and achievement. However, how pleasure, timelessness, and subjective time are related is a very tricky question. After discussing Freud and Piaget, as we will later on, we may be in a better place to proceed.

The evolutionary advances of other senses of time, particularly objective time, are matters of cultural history. To consider internal time estimation, we must first have clocks. We can define the sun and moon, their risings and goings, as primitive forms of clocks, but the kind we use so commonly today are inventions of recent centuries. And there is still the matter of subjective time perspective, our capacity to see backward and forward, in and out of ourselves. This capacity to remember, to plan, to think, may well have developed in tandem with development of our large frontal lobes. But we cannot end there. Subjective time experience does not merely engage logico-arithmetic or logico-rational capacities, but our memories of ourselves, our identities, who we are and what we live by. Able to use various measures of objective time as the track of its movement, subjective time perspective is as much a measure of affect and emotion, of what was emotionally meaningful to us in the past and, by extension, the present and future. The evolution of our various senses of time has been an activity of the evolution of our culture. It is probably unwise, and incorrect, to imagine our objective and subjective senses of time as beginning here, stopping there, in various points in history. In normal daily life, we effortlessly reconcile objective and subjective time senses. They are mutually interactive, perhaps mutually adaptive. They may well have grown and flourished concomitantly in our development as a species.

Before ending this chapter, we should add a few words on the debate regarding localizing the time sense in the cerebral cortex, a subject on which much has been written.[56] Since Sperry's work in the fifties, the whole issue of lateralization has literally taken off. Various scientists have rushed to lateralize first one, then another, brain function; others more happily lateralize brain dysfunctions.

It should come as no surprise that the two cerebral hemispheres are different, and process different subjects differently. There is probably some evolutionary advantage to both the great redundancy of brain function and the specialization on the part of both lobes; we have known for over a century that language function is almost totally lateralized in right-handed people. We now see attempts to explain the left hemisphere as mainly "sequential" in its thinking, with the right hemisphere as capable of *gestalt*, simultaneous, unitary understanding. Some have attempted to place the primary process thinking described by Freud as "probably" localizable to the right hemisphere.

These hypotheses may be somewhat premature. As of now we really do not know how the brain works, how the enormous simultaneous input of signals to the brain is processed and passed along. We are in a position to become easy victims of the fallacy of misplaced concreteness, believing our models to represent reality, when that reality is presently beyond us. The brain operates as a whole, with constant communication between parts. Given tests done in specific environments, as by Sperry's group, one part of our brain may perform a specific function. However, the next second, given a different cognitive "mix," that same function may be performed by another part of the brain. Subjective time awareness appears to be changed radically by right occipital lesions and bilateral thalatomies, while allowing the objective time sense to remain intact. Our models of how this comes about will improve as our research on the brain progresses. At this juncture we are not yet ready to claim where time is processed or lateralized, just as the great British neurologist Lord Brain wrote 20 years ago.[58]

Through the development of the neocortex, and the advance of cultural insights, humans have developed both an objective and subjective sense of time. One of the main reasons we can use all of the day as an ecological niche is that we can view time spatially, as moving backward and forward. We may consider a future we have not yet known and may never know, and a past we have not seen and yet may learn from. The evolutionary implications of our subjective time perspective are large. Most animals adapt, through their circadian clocks, to changes in the environment that are expected. The changes are internal, evolutionarily modified, and, in all probability, unconscious and unplanned. If the environment does not change as expected, function will become even more difficult.

How different we are, able to look backward into history and now, after the path breaking of eighteenth century science, into an infinite future. Humans do not just adjust innately to environmental changes; they can plan. They may anticipate, expect, consider contingency, work to provide for themselves first in one way, then in another. Through their cognitive capacities and their language, they may adapt to the future. More significantly they can adapt the future, and the future environment, to themselves.

We are not born with these capacities. We learn them through extraordinarily lengthy periods of education and apprenticeship. Born with one probable sense of subjective time, we learn to change, to adapt to our environment, and eventually to learn objective time. How we learn objective time—where it comes from and how it develops—owes much to the work of Piaget. It is the subject of the next chapter.

3
The Learning of Time—Jean Piaget

In 1929 Albert Einstein was asked to address an international conference on psychology and philosophy. Having worked out the theory of general relativity, and having already started to turn to unified field theory, Einstein addressed the audience on questions critical to his understanding of what the universe was—questions about the nature of time. Jean Piaget was in that audience. Two years earlier he had published his book *The Child's Conception of Physical Causality*.[1]

Piaget set Einstein's concerns into three main questions: Was the intuitive grasp of time innate or learned? Was it identical with the innate understanding of velocity? What bearing, if any, did these questions have on the development of the human conception of time from childhood onward? Implicit in these questions was a definition of time as it related to the physicist. The time to be considered was objective time, the time of clocks, geophysical time as humans had come to know and scientifically use it. Over the next 30 years, Piaget and his co-workers would labor on and off studying the question of time, a relatively minor component of the Piagetian edifice of the transformational capacities of the human mind, a small piece in his attempt to unify biology and logic. The consideration was always how humans learned objective time, the clock time they could use purposefully in scientific pursuits. During this period he would answer Einstein's questions in ways that would surprise many. He would declare that the human time sense is learned, and that it was essentially derived from an internalized conception of velocity. His entire conception of how the child learned time would turn on the question of velocity. He would also attempt to deny firmly the Kantian notion of time as an innate concept of the human mind, and in the process go a long way toward explaining how humans, at least in industrial societies, learn their concepts of objective time.

Piaget has described his self-constructed edifice of genetic structuralism as "very close to the spirit of Kantianism."[2] His respect for Kant was always high. Toward the end of his major work on time, *The Child's Concept of Time*, Piaget made one of his more embracing statements about his philosophical mentor, showing both reverence and a personal sense of progression and advance:

> As Kant put it so profoundly, time is not a concept, i.e., a class of objects, but a unique schema, common to all objects, or if you like, a formal object or structure. However, on the grounds that time is not a logical class, Kant argued that it is an "intuition," an "a priori form of sensibility" like space, and hence unlike the categories of understanding, e.g., unlike quantity. Now, genetic analysis has led us to a quite different conclusion, namely that time must be *constructed* into a unique schema by operations and, moreover, by the same groupings and groups as enter into the construction of logical and arithmetical forms.[3]

How did Piaget reach such a conclusion? Starting about 1930, he and his assistants inaugurated a large series of ingenious experiments with children. Though Piaget would later comment on the earlier beginnings of the child's sense of time, he worked with children who were, at their youngest, about four to four and a half years old. Insufficient linguistic development, at least for experimental purposes, appears to have been the reason for starting with such an age group (which left most theorizing on younger children's concept of time to the Freudians—see Chapter 6).

In the first of the experiments, children were presented with two flasks, one on top of the other. The two flasks were very dissimilar in shape, but held equal volumes. The top flask, of a form much like an inverted pear, was filled with fluorescein-colored water, and emptied by means of a glass tap into the bottom flask, which was perfectly cylindrical. The brightly colored liquid was allowed to flow from the top flask to the bottom, in a series of steps. The children were asked to watch the liquid as it passed to the bottom flask, and to draw the level of the liquid in both flasks at each respective level (a picture of both flasks was given to the children, who merely drew in the level on the picture). With each new level, a new sheet was handed to the children. When all the liquid had passed to the bottom flask, the children were asked to sort the drawings in order, showing how the liquid had passed down from one flask to the other.

Until about age six, they could not do it. While watching the liquid actually flow, the succession of different levels caused them little trouble. They appeared able to understand, generally with ease, that if water goes from the first level to the second, the level will go down in the first flask and rise in the second. Yet they were incapable of correctly sorting the drawings. When all that they had seen was now written down, the concept of how to arrange them properly totally eluded them. They could not describe the succession of the drawings. Until about age six, they were incapable of understanding what Piaget calls the "intuition of velocity." These children are in what Piaget calls the first stage, stage one. They cannot make sense out of "behind" or "ahead of," as applied in spatial terms. To Piaget these terms more accurately define the original conception of "before and after." Children are very capable, at age four or five, of understanding that if water goes from the first level to the second, the first level will decline and the second level rise. Yet these children cannot conceptualize that function as an idea. They are incapable of what Piaget calls "seriating what drawings they themselves had made of this process."[4]

This first step is seriation, the ability to conceive of succession. What cannot be stressed enough is that, at least for Piaget, the child learns time on the basis of first understanding velocity. As he wrote in the 1960s, "Time is a coordination of velocities, or better yet, of movements with their speeds." Even the concept of space rather rarely intrudes into Piaget's understanding of how children learn to know time.[5] For Piaget children learn how objects move—this is the fundamental understanding, the base from which they learn the sense of time. It is also at the foundation of other capabilities, like language.

For Piaget children first learn through trial and error. Then they conceptualize what they have learned. Only then will they be in a position to move on to the next "stage." Eventually children do learn how to seriate, how to order in an ordinal system, to show what came first and what came after. The next step, like all a step of discovery, is to develop some concept of simultaneity. What this really means is to learn a concept of duration. With six-year-old children and older, Piaget proceeded to another series of experiments. He placed two dolls on a table in front of the children, first one, and then the other. He would then move the dolls in front of the child, starting and stopping at the same time. But, instead of traversing the same distance, one doll was moved further than the other, though beginning and ending movement at the same moment. He thereby tried to investigate how the child conceived of simultaneity. The dolls stopped and started at the same time, yet one had gone further. Would the children be able to understand that they began and finished at the same time? Most children had no difficulty coming to the conclusion that the dolls had started at the same time, but they certainly disagreed that the two arrived simultaneously. Piaget illustrated with a sample conversation[6]:

> "When this doll stops, is the other one still moving?"
> "No, it isn't moving."
> "Then they stopped at the same time?"
> "No, they didn't stop at the same time, because that doll is ahead of this one."

Piaget says that these children cannot understand simultaneity, because they have not entered into stage two, the period of what he calls "articulated intuitions." With the advent of stage two, they will declare that the two dolls left at the same time and arrived at the same time. However, if you then continue questioning such a child, and ask which doll traveled for a shorter or a longer time, you receive a paradoxical response. The child will usually reply that one of the dolls has moved for a longer time, since it has traveled further:

> "But they stopped at the same time, didn't they?"
> "Yes, they stopped at the same time. I gave the signal and they stopped."
> "Then they moved for the same length of time?"
> "No, that one moved longer because it traveled further."

In stage two the child acquires those concepts that allow him or her to understand that the two dolls moved for exactly the same period of time. The first concept acquired is succession: the next to be understood is duration. As always for Piaget, these concepts are first understood in a sensory-motor fashion, and then, through trial and error, are gradually converted to functions that can be used intellectually. Once again all definitions of temporal concepts are in relation to velocity. For a definition of duration, Piaget states it is something inversely proportional to velocity.[7] Succession is defined by the coordinated movement of objects; his definitions flow directly from the results of his experiments, and do not consider other possibilities, other data. Thus at about age seven or eight, children are said to understand before and after. They can understand simultaneity. They can also posit understandings of succession and duration. Does Piaget consider that they thereby understand and measure time?

Certainly not. Let us quote a dialogue, here an amalgam of several rather than the record of one, between Piaget and a child of seven[8]:

> "How old are you?"
> "Seven years old."
> "Do you have a friend who is older than you?"
> "Yes, this one next to me is eight years old."
> "Very good. Which of you was born first?"
> "I don't know. I don't know when his birthday is."
> "But come on, think a little. You told me that you are seven years old and that he is eight. Now which of you was born first?"
> "You'll have to ask his mother. I can't tell you."

The critical and final period for developing an understanding of time is stage three, where the "colligation of the subjects of duration and succession" is obtained. For Piaget this is an "operational synthesis" of the two that goes far beyond an element of simple complementarity.[9] At around age eight, children have reached a position in which they develop a real understanding of time and its measurement. They have learned "reversibility."

Reversibility is something very important for Piaget. It is a fundamental of higher human thought. For him evolution is not some haphazard process that leads to a probabilistic array of outcomes. To Piaget evolution is far more purposeful; it provides for humans the ability to think in "logico-arithmetic" terms, to think as rational animals. In fact as far as Piaget is concerned, it often seems that the whole point of evolution, particularly human evolution, is this ability to think logically and abstractly.[10] Reversibility of time is so significant to Piaget because with the human ability to retrace time, both forward and backward, come memory and the concomitants of identity. The ability to see events in time, to go backward and forward with them, is a necessary construct of the logical-rational thinking style that is the human race's paramount accomplishment. Clearly Piaget will have less concern for the odd processes and vagaries of subjective time.

In the 1940s, while delineating these stages by which children learn objective time, Piaget moved beyond the concept of reversibility. In ways he did not explain except in an idiosyncratic algebraic shorthand, he saw the concepts of succession and duration, "colligatively" combined, leading to a time concept that renders time both homogeneous and continuous. Somehow, through a process of transformational operations he does not explicitly describe, human beings logically come to conclusions about time that go to the brink of a Newtonian, Euclidean universe. Through the medium of velocity, the child builds up "before and after," succession, simultaneity, and duration, via "qualitative colligations," to a system where time becomes reversible, continuous, homogeneous. It is not for nothing that the accusation has been made that "Piaget often writes as though the average person typically reasons in a way that would not disgrace a professor of formal logic."[11]

By the 1960s, Piaget was not quite as sanguine about the average human's capacity to understand time in the same abstract way in which he regarded it. During the 1940s he had gone so far as to call the operations of the time sense "infra-logical," just a notch away from the "wholly logical" or "arithmetical" thinking styles that themselves pushed toward the pinnacle of human achievement. He began to admit, in fact, that adults were often seen at the same preoperational, pre-age-seven level of children, particularly under certain circumstances. Other workers, particularly Fraisse, had gone on to show that the abstract quality of the time sense generally did not exist in youngsters until they were 15 or older.[12] It was not merely that adults were often a disappointment, but there was also the matter of the child's subjective time, out of which a sense of objective time had to be learned. This sense of lived time, of subjective time, Piaget came to see as some sort of "mistake" that individuals come to and go beyond.

Why is Piaget so quick to dismiss subjective time? One reason may be that, as he says, he regards it as an "illusion," and so not worthy of much effort in explanation.[13] Subjective time, the fact that many may feel an hour to have passed like a minute or a minute like an hour, simply does not fit the Piagetian concept of how people should think. Many of the "mistakes" that children make when learning objective time are the "errors" of subjective time. For Piaget the learning of objective time, the movement beyond a self-centered view of the world and space, is a good part of why development occurs in the first place. Humans are rational creatures, and logico-arithmetic relations are their highest calling. For Piaget, when children learn time, they learn to interpret clock time—a clock time close to physics, a time that is homogeneous and continuous, except where relativity applies. The rest is illusion, mistakes, lies of perception, the immaturity of the psyche. Even relativity may not present a problem to the Piagetian child. He applauds an attempt by a physicist and a mathematician, Abele and Malvaux, in their *Vitesse et Univers Relativiste*, to construct a relativity theory using only velocity, where time and duration do not exist at all. In fact the problems of time and time reversibility are thorny problems in physics. This simply returns us, however, to the fact that Piaget is mostly concerned with the logical, rational side of humankind. The side that is irrational, unconscious, disorderly, that takes time and turns it inside out and backward without any concern for physics, mathematics, or the real world, does not interest him very much. Humans may experience time this way; but such experiences are only "illusions."

Still, others may be concerned with such illusions, and Piaget gives some idea of how he does see subjective time in his criticism of the work of others. In some of his more recent writings on time, he attacks the work of Paul Fraisse, a friend, who postulated that subjective time is somehow related to the number of changing events in the field of observation of a subject. Fraisse himself was not entirely uncritical of Piaget's work on time—"one might say that he (Piaget) sought situations in which the relationship time–distance divided by speed was apparent."[14] Fraisse's own theory has many problems. Yet Piaget's response, following Janet, that "time = work divided by power," and that "if one increases the power, the time seems to diminish," really leads nowhere. He continues to characterize subjective time as a matter of how much energy is applied to a task. If one is interested, there will be a "mobilization of the strength of the individual . . . who whole-heartedly attacks a task important to him. On the other hand, boredom, disinterest, disassociation, can cause visible dimunition of strength, or, in other words, a shutting off of available energy."[15] Most people, however, even in simple motor tasks, can infer that energy level is not just related to whether a task is boring or interesting. And what is "energy level"? Is it the amount of ergs put out by the body? Or is it some "colligatively" derived measure of physical strength, affectual interest, pain or pleasure in a task? To assume that energy level determines the subjective duration of time really tells us little; it is the rudiment of a theory.

There are other cracks in the Piagetian system. Other workers have shown that the concept of velocity, which Piaget held to be fundamental to all understanding by humans of time, is not as close to "physical time" as he would have hoped. There are difficulties from our perceptual apparatus. For some circumstances with adults, faster equals more time. For other conditions, faster equals less time. Much of the work on velocity has been carried on by John Cohen, an English pyschologist, and his co-workers. Though agreeing with many of the experimental results of Piaget, in working with the same concerns of time, speed, and velocity, Cohen has shown that the human understanding of time is perhaps not as logical as Piaget might have hoped. What is now understood, based on work starting in the 1930s, is that time and distance are very much related to each other in human measurement. The effects of distance change estimates of time, just as changes of time influence estimates of distance.

When distance affects the estimates people make of time intervals, the phenomenon is called a kappa effect. In the 1940s and 1950s, Cohen did a series of experiments on the effects of distance on time estimation. Asking people to bisect the interval between two flashing lights by controlling the flashing of a third, he found a highly nonlinear, nonuniform effect on time perception whenever he shifted the distance between the two flashing lights not controlled by the subject.

He then repeated the experiments, but substituted different musical tones for the flashes of light. The effects were similar but weaker. He had shown that duration is markedly affected by changes in the spatial pattern of stimulation, as well as by auditory changes. Thus there were kappa effects, or effects of distance on duration, for both light and sound.[16]

Would similar effects occur in studying duration, when the changes were changes of velocity? With the advent of Sputnik, the idea of relativistic effects on time travel did not appear quite so far off. Space travelers near the speed of light would appear to age much more slowly than those proceeding at more Newtonian, earthly speeds. Cohen began to test kappa movement effects. Would longer distances appear disproportionately long? Would short distances appear disproportionately short?

Forms of ground transportation were utilized for the sake of expediency. The vehicles used were cars, which were driven throughout England. Subjects were told they would be taken on a journey during which a bell would ring. They were told nothing else. In some cases the car windows were blacked out, and in others the subjects were blindfolded. The results of the experiment showed a clear interdependence between duration, distance, and speed.[17] If a journey was split into two parts that took exactly the same clock time, the part that seemed to last longer was where distance and speed were greater. With such lack of perceptual feedback, internal time estimation could clearly be tricked.

A follower of Piaget may ask, "Were the senses of time interrelated in the manner expected, considering the physical variables of time, distance, and speed?" Symmetrical relations did occur if passengers traveled at uniform speed. However, whenever there was a change in speed, this symmetry disappeared. Passengers then generally believed that they had journeyed for a longer time after the change of speed than before it. Here, at least, Piaget's supposition about the importance of velocity in measuring time somewhat held—but not in the manner expected.

Just as there are effects of distance on time (kappa effects), there are effects of time on distance. These are called tau effects. At the

University of Nebraska in the early 1930s, Harry Helson marked off three equidistant points on a subject's forearm. He then applied mild electric shocks at all three points, but varied the interval of time between them. If the interval of time between stimuli was longer, the subject estimated that the distance between the two points was greater. Distances were changed, time intervals changed, but the tau effects remained—a longer time interval was estimated by the subject as a longer distance. Tau effects have since been replicated in a variety of subjects.[18]

Tau effects, like kappa effects, change with movement. In the 1960s, Cohen marked off a short course of about 150 yards. He then asked his subjects to walk the first half and run the latter half, without telling them that the marker was placed exactly at the halfway point. The results showed two divergent responses. Almost half the group felt the distance was greater when walking than when running; the rest thought it greater when running. The magnitude of effect was greatest for those who claimed that walking made the distance appear longer. Results broke down almost evenly between those who felt distance increased with a longer time interval and those who felt distance increased with a shorter time interval. Here was a double perceptual "misreading" of the Piagetian principle of velocity defining time, as well as a highly inaccurate estimate of two equal distances as unequal.[19]

These experiments show that the capacity for internal time estimation, the innate human capacity to estimate time, is affected by distance, light, and auditory effects. Yet our sense of objective time, particularly internal time estimation, is even more complex than that. Recent work has shown the major importance of our concept of space.

Spatial scale strongly affects our sense of time. A few years ago, students at the University of Tennessee were asked to walk through a series of constructed lounges. What made the lounges particularly odd was that all had been made to scale—cardboard partitions, chipboard furniture, replicas of human figures. The lounges were made to 1/6, 1/12, and 1/24 of real size. The students were asked to move around the lounges, familiarize themselves with them, and finally imagine themselves a part of the environment. They were then asked, at the various scale models, when they subjectively felt (not thought) that 30 minutes had passed.

Here is one of the relatively rare studies where normal people have been asked to estimate their own subjective time. A variety of interesting results came out of the study. In the realm of subjective time, subjects in the environment of 1/24 scale estimated 30 minutes to feel about 90 seconds long. At 1/12 scale 30 minutes were estimated at around three minutes, and at the 1/6-scale lounge, 30 minutes were felt to be over when about five and a half minutes of real clock time had expired. Taken to a regular environment, scaled one to one, and left without cues, subjects estimated 30 minutes to be about 30 minutes. Thus time, as experienced subjectively, worked out almost in exact ratio to the experienced scale of environment, at least in this experiment. The subjective time duration in the 1/6-scale environment was about twice as long as that of the 1/12-scale, and 1/12 was about twice as long as 1/24. All time experiences were radically different from what would have been estimated by the same subjects if they had been asked to estimate objective time in a normal environment.

A curious subgroup developed. Some of the students did not follow the instruction to tell how much time they "felt" had passed. They suspected the experiment was part of some "test," and decided to use internal cues to estimate the time experienced. In other words, asked to record their subjective time durations, these students responded with verbal estimates of internal time estimation. Asked to give an estimate of how much time they *felt* had passed, they estimated how much time they thought had gone by on an objective, external clock.

This group, particularly when masked to all but the miniature environments, still gave estimates of objective time that were sharply different. When asked to estimate 30 minutes at 1/24 scale, their mean score was about six minutes, which increased to seven and a half minutes when asked at 1/12 scale. It should be remembered that individuals at normal scale gave internal time estimates very close to 30 minutes.

This work by DeLong has provided some useful quantitative findings. First, in similar experimental conditions, subjective time was very different from objective time. The time we feel and the time we know are very different. Just as interestingly, both internal time estimation and subjective time duration have been shown to respond strongly to spatial scale. This in itself is not an entirely unexpected finding. Many stimuli that affect objective time affect subjective time; but how, and in what direction, has rarely been studied.[20]

As we return to Piaget, we must modify his consideration of velocity as it relates to time. Though it may have a great deal to do with how humans learn to measure time, velocity is not the sole measure we use. There are effects of velocity on our objective sense of time, but there are also potent effects of distance, spatial scale, and visual and auditory stimuli. And there are still others, as we describe in Chapter 5. The human sense of time, both objective and subjective time, is clearly related to a large number of factors in the external and internal environment.

Piaget's relative indifference to subjective psychological time should not obscure his major importance in helping us to understand how humans learn their sense of time. He did not just argue about the Kantian conception of time as an "innate" quality. He studied it. He showed how much of the human sense of objective time is learned, and provided stages to describe this learning. He showed how these stages produced progressively different conceptions, from one to the next, of time and memory. He was also able, later on, to modify his thinking slightly, and accept that adults often return to "preoperational" stages of experiencing time. Though in retrospect Piaget was not totally right, he was often so. He began a course of study of human development and of how the mind works that is ongoing and flourishing, and pointed out some major factors in that development, if not all. One of the factors that Piaget did not consider much, working so exclusively with Genevan schoolchildren, was the overall importance of culture. Throughout the world there are still great masses of people who have never encountered Piagetian "formal operations," and exist in a primarily preoperational world. Do they experience objective and subjective time as we do? That is the subject of the next chapter.

4
Time in Different Cultures

Over the past 50 years, a great deal has been learned about how children in westernized, industrial societies learn to tell time and internally estimate time. We know far less about subjective time, and this partially may be the fault of Piaget, who regarded it as an "illusion" corrected by the learning processes of childhood. For him the fact that time is somehow not always placeable in an objective, overarching framework was rather objectionable, for it did not quite fit his conception of human beings as logico-rational creatures. Still, having some idea of how objective time is developed in normal children, one can consider how culture, mental illness, and personality might affect both objective and subjective time.

What about people without clocks? The ready availability of clocks, even in western industrialized societies, has made headway only since the fourteenth century. To a great extent, the mass production of clocks and watches, and their easy availability to most people, has existed only since the nineteenth century. This development might pose a major historical difficulty with Piaget's theory, particularly as it relates to time. If formal operations began with the Greeks, was that also true of their capacity to use and tell time? What about the period when societies had no cultural ability to use formal operations at all? We might find ourselves in a historical bind, at least theoretically, in understanding how time has been learned and used, within the larger, embracing Piagetian view of things.

Yet we are not historically restricted in attempting to understand time conceptions that predate clocks. Many societies still exist to whom formal operations are not only a foreign concept, but in many ways an unintelligible one. And though telecommunications and technology are advancing far more rapidly than formerly, there is now a historical record to look to.

With the advance of nineteenth century imperialism, and its subjugation of "primitive" peoples, came the concomitant advance of ethnography. Though at the beginning, ethnographers were often employed by colonial authorities whose main interest was more effective control of native populations, they soon created an academic discipline that attempted to understand and preserve knowledge of these cultures. Most ethnographers soon learned that the word "primitive" referred mainly to the technological state of the people with whom they lived. They generally discovered social and political organizations with complexities similar to those at home, though differently placed, usually in kinship lines and nuances of social ritual. The word "primitive" has risen and fallen in appreciation through the decades of anthropological literature.[1] The majority of long-term field workers came to believe that it did not possess a pejorative connotation, but described instead a richness of tradition and conceptualization that those from the West could only partially understand.

What does time mean to a people without clocks? The anthropological literature does not offer a great number of systematic studies of time sense. Those that exist, however, were carried out by some of the best ethnographers of the structuralist/functionalist school, such as Evans-Pritchard and Bohannon.[2-4] This relative lack of interest in time as a subject of study is not very surprising. Most interest in the subject arises in the psychological community, and there it is often greatest among psychophysiologists. Perhaps for reasons of convenience, and perhaps for reasons involving experimental design and simplicity, psychophysiologists have tended to disregard the possible effects of culture. Historians and social scientists, though the situation is changing, have usually shown little interest in the subject of time, and in the case of anthropologists, we can easily understand why. To know the time sense of a people, particularly how they use their subjective time sense in daily life, is to know those people as if one had grown up with them. More significantly, in societies that are ordered by tradition and larger religious-political-social frameworks, it is often almost impossible to divorce time conceptions from the larger ideas that govern societal behavior.

One of the finest accounts of time in a people without clocks, and who cannot even consider the concept of clock time, concerns the Tiv, an agricultural people of central Nigeria. Conceivably things may have changed among them recently, as they are now citizens of a rapidly modernizing, if chaotic, nation. Yet in the late 1940s and early 1950s when Bohannon lived with them, the inroads of industrial civilization were few,[5] and we will base our discussion on the conditions of that era.

To Bohannon the Tiv do not so much measure time, as indicate its movement. Like many other primitive societies,[6] there is no equivalent in their language for the word "time." Although there are several words that express differing senses of duration, and others that mean occasion, there is no word equivalent to our "when." Placing an event in time is generally done by referring to a condition of natural or social life, and associating one event with another. Association and causality are very different things, as we will show for the Tiv concept of time.

Though lacking a calendar, the Tiv have a five-day week, and count "moons" and "dry seasons." These may appear to us, at least superficially, as equivalents of months and years. In reality they are not. The day is separated into day and night. There are far more names for parts of the day than for parts of the night; to describe a time during daylight, the Tiv generally point their fingers to a position in the sky that the sun will then occupy. Some hint of the symbolism of Tiv meanings can be garnered by knowing that night is far less specifically catalogued, partly because of what is said to occur during it. The word for "daylight" also means "not connected with witchcraft." Witchcraft is for the night.

Days of the week do not follow geophysical absolutes, but are named by market town. They are not absolute concepts and do not have names as such, but follow distinctly local custom. If you are in the Iyon district, the second market day is Iyon. In South Ute the same day is called Pev. Day two in one district is not "day two" in the next. The Tiv will understand that today is Iyon in Iyon, and that the same day may be Pev in South Ute, but that is not important to them. Where you are determines what you call the day. If you are in South Ute, the second day is Pev, regardless of what it is called ten miles away.

The fact that time is defined by social custom in association with natural conditions is itself defined by Tiv concerns for season. They have no word for season, but divide the year into many different sections, depending on rainfall, corn planting, winds, and so on. When describing one of their northerly winds and its appearance, the Tiv sometimes say, "We cut the guinea corn when the first harmattan comes." However, they just as easily say, "The first harmattan comes when we cut the guinea corn." The two statements are considered equal. One event does not cause the other; they are merely associated. According to Bohannon no causality at all is implied in the association.

What this means for absolute time, or clock time as we understand it, becomes apparent when you ask a Tiv how many moons there are in a year. The answer is something between ten and 18. How many days in a moon? The answer: between ten and 50. Accuracy of time keeping, as we understand it, does not concern the Tiv. Women, when pregnant, count the "moons" until their childbirth. The Tiv consider that human gestation takes nine moons for a male and eight moons for a female. Women mark the number of moons on their huts. The marks generally are not the expected number when the baby arrives, but this is imputed to human error. Similarly, males are organized into age sets that are formed about every three years. Bohannon discovered that all members of a set nevertheless considered themselves to have been born in the same year. He pointed out this discrepancy to the Tiv. They admitted he was "quite right." Nonetheless they were, as they told him again, born in the same year.

To the Tiv, as among the Nuer, history is relatively shallow. Genealogies do not go back very far, and the concept of "long ago" may refer to an event an hour past or something much more distant; "long ago" defines not so much an event in time as something established or traditional. "Long, long ago" describes an event in a dim past, a past with a quality as vague and unknown as that of the future. The word for future translates as something like "in front." When pressed, the Tiv can easily adapt their lineages into foreign historical contexts. Hearing the Europeans tell their own creation myths, they easily accepted "Adam and Ife," the latter a very Nigerian version of Eve, into their lineages.

We may ask what happened when the Tiv finally became acquainted with clocks. With regard to European time pieces, some Tiv soon came to use the word "ahwa" for "hour." But this was a word similar to their word for "mark." One o'clock was "one mark," five o'clock was "five marks." Bohannon was convinced that the Tiv had no conception that watches marked off units of time. It was simply a mechanical European object of moving parts, with no apparent use.

Thus, for the clockless Tiv, time has a very different meaning than for us. That is not to say, however, that the Tiv are without objective reference points. They certainly possess them, but they are more overtly social than geophysical. Time for the Tiv is also not absolute, continuous, or homogeneous, as objective time is for us. Bohannon asks a question Piaget might ask. Can the Tiv "colligate duration and succession"? Can they place events in a time series, like a man's marriage and maturation rituals, in specific relation to other events, such as the movement of another individuals' age sets? Bohannon states that this was a problem for experimental psychologists, who might usefully apply themselves to studying the question. Unfortunately the experimental psychologists went to other places for their studies, and sometimes left their anthropological books at home. We still do not have an answer, as we will see later when we consider their results.

The Tiv live without an absolute time, a clock time. But it is perhaps unfair of Bohannon to say that they simply "indicate" time, rather than measure it. They do measure it, when it is socially meaningful for them to do so. To our concepts of social meaning and social time, however, they are relatively immune. Perhaps we do better to consider that our concept of absolute time, so necessary to our science and daily life, is as much a cultural element as the Tiv conception of a female birth taking eight "moons" versus nine for the male. Our lack of historical sense in daily life often allows us to forget the theological and social roots of our own concepts of time, though they may appear obvious to us in the case of the Tiv. Indeed Tiv conceptions of time may strike us as far closer to subjective time than as concepts of objective time.

Such a conjecture would be a mistake. The Tiv sense of time, of harmattans, and moons, and weeks, is not private but social; it is not personal, but embraces the entire society. It is as much a part of a unified system of living and believing as our own view of objective time. There are probably as many different objective "times" as there are cultures. With every individual these times will be uniquely and continuously reinterpreted. Objective time exists for the Tiv, but it is very different from our objective time, at least as it relates to clock time. To say that the Tiv are "timeless" or "cannot keep time" is incorrect.

Nevertheless, the vast differences in time keeping between primitive and industrial cultures has led to a great deal of theorizing on the part of scholars from the latter. Mircea Eliade is very frequently cited in discussions of "primitives," particularly in reference to his conception of the "myth of eternal return." For Eliade many primitive cultures possess a cyclical view of life. What was the past becomes the present, and will be the future. Through ritual, the primitive "nullifies time," and forgets the uneasiness of death. The primitive's goal is to live in an "eternal present." For Eliade the eternal present is a point where:

> Everything begins over again at its commencement every instant. The past is but a prefiguration of the future. No event is irreversible and no transformation final. In a certain sense, it is even possible to say that nothing new happens in the world, for everything is but the repetition of the same primordial archetypes; this repetition, by actualizing the mythical moment when the archetypal gesture was revealed, constantly maintains the world in the same oral instant of the beginnings. Time but makes possible the appearance and existence of things. It has no final influence upon their existence, since it is itself constantly regenerated.[7]

Though some societies, particularly some Australian aboriginal tribes, may show evidence of conceptual systems that are cyclical and endless, as during their "dream time,"[8] they are probably the exceptions. Eliade's conception of time in "primitive societies" may lie more in Western mystical thought than in ethnographic evidence.

No less an anthropologist than Edmund Leach has responded to this enormous, encompassing conceptual scheme of primitive time. He sees primitive time as far more discontinuous. The past is not so much filled with content, as shallow with it:

> Indeed in some primitive societies it would seem that the time process is not experienced as a "succession of epochal duration" at all; there is no sense of going on and in the same direction, or round and round the same wheel. On the contrary, time is experienced as something discontinuous, a repetition of repeated reversal, a sequence of oscillations between polar opposites; night and day, winter and summer, drought and flood, age and youth, life and death. In such a scheme, the past has no depth to it; all past is equally past; it is simply the opposite of now.[9]

But this is a special Leach. This is a Leach falling under the spell of Levi-Strauss, a spell that was relatively short-lived. The polarity and binary structure of symbol systems are things Leach came to eschew later in his career. The simplification this demands in explaining any life or society has fallen into increasing disfavor. Clock time may not exist in primitive societies, but this is not to say that objective time does not exist. It is measured, and defined, in social and natural terms, and though it may not have the precision of our geophysical and absolute clocks, objective time is still apparent in these societies. We sometimes read statements such as Kluckhohn's in reference to the Mexican villager: "He lives in a timeless present. Tomorrow in a highly specific sense is meaningless to the Spanish-American or Mexican."[10] But such statements are far less common today in the anthropological literature.

Some of the confusion about objective time in primitive society may lie in linguistic analyses of time sense in these other cultures. Benjamin Whorf pointed out that the Navaho do not have words for "past" or "future," and some took this to mean that the concepts were lacking. But does the lack of "past" and "future" as separate words really mean that the Navaho, or any other people, do not conceive of a past or future?

One typical proponent of this linguistic determination of reality is Hajime Nakamura. In describing Indian civilization, he places great reliance on language as obtaining very wide psychological significance:

> Language, as usual, is where this lack of common sense (time) concepts is mostly clearly seen: The Indian people did not have clear awareness of the discrimination of tense, although in Sanskrit, as in Greek, there are five kinds of tenses that are not sharply discriminated in meaning. To indicate past time, the imperfect, perfect, past participle active, aorist and historical present are used almost indiscriminately, and the frequency with which a given tense is used varies not according to meaning but according to historical period.[11]

From this consideration of tenses, Nakamura concludes that Indians never attempted to grasp a quantitative sense of time, and never wrote histories with accurate dates.

Is lack of tense always psychologically significant? English lacks a subjunctive tense, yet is quite capable of expressing what other languages do with the subjunctive. One interesting vignette, noted by Doob,[12] concerns an African theologian describing the difficulties of teaching eschatology in Kikuyu. Despite its three future tenses, Kikuyu only considers actions occurring two to six months from now—the distant future simply does not have a tense. How then does one explain the infinities of the Old Testament to the approaching faithful? But does the lack of a "distant future" tense make such a future inexplicable, or does it simply make it more complicated to explain?

It may be that some of the writing on time and language in the anthropological literature is more a comment on the dangers of translation than on differing concepts of time. Clearly, among the traditional Tiv, as among the traditional Navaho and the traditional Kikuyu, time has very different social meanings. The future and the past mean different things than they do to us—but that does not mean that they do not exist. It is more correct to say that time, despite linguistic differences, does exist as a concept in primitive societies. It also involves objective features, as it does in our own. However, the meaning of time shifts, often markedly, from culture to culture.

We may see a less extreme example of the difference of meaning of clock time in studies done in Brazil. In a comparison of Brazilians and Americans, Brazilian adults seem to possess a more "flexible" view of clock time, at least in social settings, than their northern counterparts.[13] In one study an objective measure of "objective time" was undertaken. Watchless Brazilians, personal watches, and public clocks were all less accurate geophysical timekeepers than similar forms of measurement in the United States. When asked the time by bystanders, Brazilians gave less accurate renderings, and appeared to define "early" and "late" for social engagements more leniently than did Americans. Even with clocks, objective time may be more or less "objective," depending on who reads the timepieces. Clock time means different things in different places. Through personal experience we also know it means different things to different individuals.

Let us now turn our attention to internal time estimation—the ability to estimate, without clocks, the passing of objective time. When studies of internal time estimation were finally carried out with clockless people, it was expected, as a matter of course, that primitive peoples would be highly inaccurate. After all, if we are to believe Eliade, and even Bohannon, clock time is a very different, very alien concept for most primitive peoples. Often they are not privy to such a view of time at all, and possess nothing similar to a Western concept of time. So how could estimating the movements of a second hand mean anything to them, particularly where they have no concept even remotely similar to a second? The results, rare, and even more rarely looked at, were not as expected.

During the 1960s there was a great interest in the changes evoked by "modernization" throughout the third world. One of the more interesting studies of modernization was carried out among the Kpelle people of central Liberia by Gay and Cole.[14] American Peace Corps volunteers had just been sent to the agricultural Kpelle when the research was carried out.

Kpelle concepts of time have a great deal in common with those of the Tiv. Though the Kpelle might possess a more specific idea for year than do the Tiv, they have no word for time. As with the Tiv, time of day is marked by pointing at a position of the sun in the sky. Months, rarely counted, are never counted in numbers higher than three. The age of people is not known with any yearly distinction beyond the age of four or five. The only years that are marked in people's memories are the years when someone is born, joins school, and dies. The years between are not counted.

Along with having no word for time, the Kpelle have no words for short periods of time. Seconds and minutes are units of time that are meaningless to them, and these are the units of classical use in psychological studies of internal time estimation. The Kpelle neither measure such units, nor indicate them. The authors fully expected Kpelle subjects to be more inaccurate than the control group of Westerners. After all, they had no concept of what was being studied in the first place.

How do you test a Kpelle, and have him or her estimate time, when he or she does not know what a short time measure is or means? Two tests, concrete in style but rather intelligently devised, were done with the Kpelle, with both children and illiterate adults. In the first test of time estimation, the Kpelle were asked to pace off various distances—20, 40, 60, and 80 yards. As they paced off these distances, they were timed. Then they were asked to pace off the same distances "mentally". When they had "mentally" finished, they told the examiner, who timed these mental estimates.

In the other test, the Kpelle were shown a stopwatch, and were asked to observe it tick off various periods of time. A variety of intervals, ranging from 15 seconds to two minutes, were used. The examiner then held the stopwatch in a place where the Kpelle could not see it. After he had shown them one of the intervals, he would start the watch away from their view, and ask the Kpelle to tell him when the second hand reached the same place as just before. The same two tasks were later performed by American adults.

On the first task, pacing off a distance first physically and then mentally, both Kpelle schoolchildren and illiterate adults were considerably more accurate than the American adults. On the second task, Kpelle schoolchildren were most accurate, with both adult groups about equal. The variance of the Kpelle groups was greater than that for the Americans, particularly on the second task. The examiners had no idea how the Kpelle managed to be so accurate in internally estimating clock time, particularly as they lived without clocks. They suggested that beliefs that Africans did not possess a sense of time were "nonsense."

Following up Gay and Cole's work, Robbins, Kilbride, and Bukenya performed the same tests with rural and urban Bagandans of Uganda.[15] They wanted to know if the rural Bagandans, living without clocks, would be less accurate than their urbanized, more westernized fellow tribesmen. To their surprise, both groups gave very similar results, when estimating 15-, 30-, and 60-second intervals. The results were, if anything, more accurate than many Westerners attain. The rural group did, however, show greater variance of results.

There is not a great deal of work on the cross-cultural attributes of internal time estimation, but if we can abstract somewhat from what data there are, we may have evidence for general human similarity when estimating internal time. Given explicative circumstances, "primitives" lacking clocks were every bit as able to estimate time internally as the average American undergraduate. Perhaps the cognitive and biological underpinnings that allow us to estimate time internally are so basic to our lives and actions that they exist almost universally.

If we try to study subjective time, particularly subjective time duration (our sense of time passing slowly or quickly), we find ourselves in a methodologic bind. Even in psychology and psychiatry, subjective time is rarely studied, perhaps because of the difficulties in quantitating it.[16] It should come as no surprise that subjective time duration is equally unstudied throughout the social sciences. Virtually the only evidence we have comes from one writer, who notes that wherever he has studied time, whether in Africa, Asia, Europe, or America, people tell him that time differentially passes slowly or quickly.[17] The sense is that subjective time duration changes may exist everywhere.

The whole concept of subjective time demands that we understand a culture in detail, as subjective time is intimately connected with concepts of cultural and personal meaning. We know that subjective time varies greatly for us, depending on whether we are bored or interested, happy or unhappy, doing something we like or something we dislike. To understand that on the cultural level may be much more difficult than it first appears. How do we know what is innately exciting to the Tiv people? How do we know what is of great psychological import to such a person at any given moment, and how it relates to that person's cultural conceptions?

The problem is by no means impossible. Yet we must remember that we ourselves infrequently consider the concept of subjective time, at least in a conscious fashion. The fact that we feel a time period to be lengthy when it was, in real-clock terms, quite short, is an element we easily and most often subconsciously integrate with our objective sense of time. The discrepancy generally does not bother us.

In the eighteenth century, the Italian philosopher Giambattista Vico expressed the belief that man would achieve greatest understanding not in the physical or natural sciences, but in the human sciences. A man could not think or act like a rock or a bird, but he could think or act like a man, and understand another man, because he was one himself. To understand history, and to understand other people, was to recreate their understandings, to recreate their feelings and hopes and ideas in the mind of the scientist. It will remain a subject for philosophical debate as to whether understanding is more possible, or complete, in history than in physics. In our century a concept similar to Vico's is that of the great sociologist Max Weber when he speaks of *verstehen*. In literal German, *verstehen* means understanding. For Weber *verstehen* as used in sociology has a different meaning, something akin to "sympathetic understanding." It is the ability to get inside the bodies and minds of the people being studied in an attempt to understand their behavior.

To a large extent, concepts such as *verstehen* must be utilized if any understanding of the subjective time senses of people of other cultures is to be attempted. Even more than with subjective time duration, this is particularly difficult in trying to understand subjective time perspective—how an individual views and values the future and the past, and how they are connected.

It is perhaps not difficult to ask primitive people, or others, if, during some incident, they find time to be passing slowly or quickly. It is a very different matter to understand someone's description of the future and the past, particularly when that individual's culture is radically different from those in the West. As a result the forays of social psychological inquiry into the future senses of nonindustrialized, primitive peoples have generally afforded results that say somewhat more about the researchers than about the people being studied.

Many of the studies aimed at understanding the future orientations of so-called primitive peoples have tried to do so through the methodologic back doors of "achievement orientation" and "delayed gratification." The assumptions of such studies deserve at least preliminary examination. In the West "everyone knows" that people who excel at delayed gratification are "more oriented toward the future," as they can wait to obtain what presumably will be greater rewards. Similarly, as these studies imply or state, those in our society who are "achievement oriented" are almost always more "future oriented" as well.[18] It was thus surmised that psychological tests of future orientation and achievement orientation could be utilized to aid cross-cultural study and comparison. One reviewer of the cross-cultural "delayed gratification" literature mainly agrees with the reported results—the less industrialized a people, the less able they are to delay gratification.[19] Such psychological studies have often left out, or have not looked at, the cultural implications of what was being studied to the people being asked. What does how long you can delay yourself from eating candy say about your "future orientation"? Let us review the methodologies of some examples.

In the 1960s an attempt was made in Tanzania to compare two peoples who appeared to have very different considerations of success, at least as defined in the West. The Chagga, a successful farming people of the Kilimanjaro, were compared with the more "laggard" Bondei Moslems of the coast. The Bondei's income and industriousness were far outstripped by that of the Chagga. The points of comparison were made through psychological tests of achievement and future orientation, obtained from the respective groups' schoolchildren.[20]

When Ostheimer did these studies, the tests he used had been devised in the United States by Robert Knapp and his colleagues. When studying American undergraduates, Knapp had noted that "achievement orientation" correlated well with the degree to which these students preferred the colors of various Scottish tartans. Those who were "success orientated" seemed to prefer somber tartans, particularly those with blue or green, over more "active" red tartans. [21] We should note that "success orientation" and "achievement orientation" were measured by scores on the Thematic Apperception Test. Thus was born the "tartan test," to be used in several studies throughout the world. Correlation somehow equated with causality, Knapp argued that the color preferences of undergraduates clearly said something important about those who wanted to succeed. The people who preferred somber colors did so because they wanted a soft environment. They wanted a soft environment in order to manipulate it, eschewing "strong, intrusive stimuli," such as reds. Thus color related to achievement orientation related to future orientation.

Along with the tartan test, Knapp went on to devise the time metaphor test. Working with undergraduates, Knapp independently devised 25 metaphoric "images." He then gave the images to the undergraduates, and asked them to order them in accordance with how much each image satisfied their "personal sense of time." Some images were fast moving—"a dashing waterfall," "a galloping horseman." Others were intentionally slow, such as "a stairway leading upward" or "a quiet, motionless ocean." Once again Knapp found correlations with the achievement profiles on the TAT. The "achievement oriented" preferred images of swift movement. He also felt that those who were not so oriented preferred slow images.

Knapp did not go on to explain why a "spaceship in flight," the swiftest object mentioned in the list given to the undergraduates, had a zero correlation with achievement scores. He also did not explain how "a devouring monster" was presumably the slowest image, as it possessed the greatest negative correlation. The fact that the "rock of Gibraltar" had one of the highest correlations with achievement, despite its apparent "slowness" as a metaphor, he ascribed to the advertising campaign of an insurance company, neglecting its more general cultural meaning. In no case were the achievement orientation scores of the subjects compared with the real article.

Ostheimer gave both of these tests to the two groups of Tanzanian schoolchildren. Perhaps unsurprisingly, both the time metaphor test and the tartan test of achievement orientation and future orientation led to inconclusive results. In the words of the examiner, "This research does not encourage future achievement testing in Tanzania with the instruments here employed."

We must use our imaginations in picturing how the farm children of the Kilimanjaro interpreted Scottish tartans. We can also only speculate as to how these children, hundreds of miles from the sea, interpreted Knapp's metaphors of the time metaphor test. What did they think of in the 1960s when they read "a spaceship in flight," "a quiet, motionless ocean," or "a speeding train"? We can only imagine how they imagined them, if at all. Projective tests, even in Western populations, are of doubtful use in explaining people's future orientation, but their use with non-Western peoples is even less warranted.

In a somewhat similar vein, it is interesting to consider an Australian study of delayed gratification among the Australian aborigines. In reviewing the international literature, Bochner and David discovered that delayed gratification tended to increase in children as they aged.[22] Similarly, a greater capacity for delayed gratification appeared earlier in those who were "intelligent," at least intelligent as defined by Porteous maze tests used to determine I.Q. They expected similar results among the Australian aboriginal children. The methods of the study were of the general form of many such studies—children were given the choice between a smaller amount of candy or money now, and a larger amount of either one week hence.

The researchers were rather surprised when, unlike the expected result, there was no correlation between the ages of children and delayed gratification, methodologically defined. Even more surprising, the correlation with I.Q., as measured by the Porteous maze test, correlated negatively with the ability to "delay" gratification. Those who wanted their candy now had I.Q. scores 13 points higher than those who could wait. The author's explanation was as follows:

> The positive relationship between immediate gratification and intelligence found in the present study could be a function of the extreme harshness of the Central Australian environment, where, paradoxical though it may seem, there exists such a sparcity of food that foresight just cannot operate, and the more intelligent action is to take what is offered when it is available.[23]

Is it true that "foresight just cannot operate" for the Australian aborigine? Such conclusions are a disservice. The authors apparently did not ask the aborigines their opinion. They have little knowledge as to what was going on in the heads of the children to whom they gave candy, and they have no knowledge of what such "delayed gratification" means to these children. Given the circumstances of the trial, almost any interpretation is possible, and their relative likelihood is no better than a random guess. Higher Porteous maze scores do not make "immediate gratification" more intelligent or less intelligent in any particular environment. It is far more probable that the concept of "delayed gratification," particularly as it relates to "future orientation," holds very different meanings and attributes for the aborigines.

What these and other studies show is not necessarily that it is difficult to study subjective time perspective among nonindustrial peoples, but that to do so demands a full understanding of how they work and think. The concepts of both objective time and subjective time, particularly when considered in their subclassifications, vary greatly from culture to culture. If we can hazard a guess, it may be that subjective time, more intimately concerned with affect and meaning, may possess the greater variation, particularly when technological components of societies are relatively equivalent. But we will now move from societal definitions of time to more personal ones. Much of the work on different senses of time has been done with the mentally ill. For reasons that at present can only be hypothesized, time senses seem to vary widely across various mental conditions. How these time senses vary, differently with different kinds of illnesses, is what we discuss in the next chapter.

5
Time and Mental Illness

Time has different meanings from culture to culture, from period to period, and from person to person. From the standpoint of trying to place time in some sort of psychological perspective, we encounter another difficulty. One of the more alarming aspects of time research is that, to a sometimes major degree, time has a different meaning from study to study.

Methodological problems are always rife in any field of knowledge, and particularly in a young field of knowledge. Psychology and psychiatry have taken a large share of blame for the research procedures they use. Such concerns are inevitable—you have to know if you are really measuring what you think you are measuring. The methodological problems of studying time are compounded in studying the mentally ill. The study of time, with its elusive, differing cultural meanings, and vast array of measuring forms, is difficult enough with so-called normals. It should be expected that the procedures used in working with normals may not readily adapt to populations whose cognitive structures are often, at least temporarily, very different.

One must also confront the importance of diagnosis. First, there is the issue of validity. In looking at the historical progression of what are now three diagnostic and statistical manuals for psychiatry, we know that what we call schizophrenia today is not what was called schizophrenia ten or 15 years ago. It will probably be something different ten years from today, when we will have the impact of a fourth *Diagnostic and Statistical Manual*. The definitions of other illnesses no doubt will also change. This is highly problematic in reviewing the literature on time and the mentally ill, much of which is old.

Next there is the issue of reliability, or the sense that what one researcher called illness *X* would indeed be called illness *X* by another researcher. This problem is again particularly daunting when one looks at the older literature, where the criteria of diagnosis generally are not stated. When the criteria are stated, they are often of the type, "mutually agreed upon diagnosis." It is as if the diagnosis of "schizophrenia" had some overwhelming, absolute meaning.

In looking at what has been written about time and mental illness, we must also confront trait versus state arguments that are endemic in psychology and psychiatry, and find out if we are really dealing with more or less homogeneous populations. This desire for homogeneity of population is not frivolous, but lies at the base of the scientific method. If a population is not homogeneous, our results may be hopelessly biased, in directions that, a priori, we cannot name. The statistical techniques we use so frequently and casually are inapplicable if the groups studied are not homogeneous for the conditions we wish to study.[1]

In the work that has been done on time, the populations examined, like those in acute inpatient hospital units, are often not homogeneous. Diagnostic criteria vary from study to study. The methods of measurement, and the forms of intervention, are different. Despite the difficulties, details do stand out; commonalities appear, and what at first glance seems to be overt contradiction, may represent the differences between two populations. What is important is cautiously to look over all studies, with particular interest in how they came to their conclusions. To count the numbers pro and con, on any single issue, provides little information. Often claims are made, especially in the literature regarding subjective time perspective, that are not supported even by the prima facie evidence itself. Some studies are better than others, and the points of difference demand consideration.

The voluminous literature of time and mental illness, despite its size, has concentrated on a few select conditions. Within the literature available, there is enough on three topics,—schizophrenia, affective disease, and "antisocial personality"—to permit some generalizations. This chapter is divided into four sections, dealing with these three conditions, as well as what might be called the strange case of normals. I will try to classify what is usefully known by reference to the classifications used throughout this book: external time calculation, or clock time; internal time estimation, which is how well people estimate time without clocks; subjective time duration, or how slow or fast the present feels; and subjective time perspective, the affective and cognitive placement of past, present, and future. Unfortunately, there is often little known or studied about subjective time, and the two subclassifications of subjective time may need on occasion to be combined.

All generalizations about time and mental illness must be made cautiously, and the generalizations made here will need to be looked at in such a light. If more facts could be known, it seems that the traditional psychoses represent a fertile field for the study of subjective time, as opposed to objective time. It appears that there is an unusual human capacity, perhaps lodged in our biological clocks, perhaps in our cognitive skills and memory, that allows to all but the most ill the ability to read and estimate objective time. Subjective time, with its inherent uncanniness and response to emotion, may be the time sense where mental illness first becomes pronounced and, later, most obvious. The pattern of that subjective time distortion, unlike that found in "normal life," may be relatively specific, at least in affective disease.

SCHIZOPHRENIA

External Time Estimation (Clock Time)

Most studies done with normals never question the ability of subjects to "read" a clock correctly, or to answer questions about day, month, and year. This assumption may not be wholly acceptable, as some New Jersey neurologists discovered when they asked normal people basic questions about time orientation. A full quarter of those interviewed, who had a high school education or less, inaccurately gave the day of the month, or did not know what it was.[2] People with higher educational backgrounds gave only slightly more accurate responses.

The problem of time orientation in schizophrenics is much more difficult. First, what is schizophrenia? In the past few years, the line between affective illness and schizophrenia has blurred yet further, and differentiation of these two difficult-to-define syndromes is a major task of future diagnostic classification.[3] Bleuler considered a clear sensorium a hallmark of his definition of the schizophrenic syndromes, but there has always been some question about the validity of that yardstick when looking at those who are acutely ill. Here the distinction between affective and schizophrenic syndromes becomes yet more difficult to determine. The European predilection for "acute reactive psychosis" is generally not acceptable to American clinicians, who usually try to define whether an acute outbreak is a variant of schizophrenia, an affective disease, or something different. The similarity of response to different medications, themselves rather gross modulators of neurohormonal regulation, does not aid matters. Even when looking at chronic schizophrenia, we have to ask ourselves precisely what diagnosis we are really talking about. We can probably put the question differently, by asking whether the diagnosis of schizophrenia describes individuals with very similar signs, symptoms, and cognitive functions.

With regard to time, and the ability to read clocks and do the rudimentary elements of a mental status examination, the answer is no. In the 1950s investigators in New York State were surprised to find multiple examples of a clinical state that was previously known only anecdotally. Some of their schizophrenic patients, all of whom had been hospitalized for many years, claimed that for them time had literally "stood still." In Germany in the 1920s, Fischer had described one such patient:

> Previously I was a human being with body and soul and now I am not such a being. I don't know anything anymore . . . The body is light and I am afraid it will soon fly away. I continue to live in eternity. There is no hour, no moon, no night . . . Time does not move. I am really between past and future.[4]

Some of the New York patients went further; "I know I should not have white hair since I am only 28 years old." The man was really in his 60s. Lanzkron and Wolfson selected a series of these long-term patients. Most of the them knew their date of birth, but were wildly off when asked their present age. As a group they claimed an age of about 26; their actual ages averaged over 47. Their real ages when they first came to the hospital were within six months of the "current age" each had supplied the examiners. The authors saw this as indicating that such patients felt time stopped moving forward six months after they entered the hospital.[5] Dahl, doing a follow-up study with another state hospital population, found a sizable percentage of his patients thinking similarly. They knew the year of their birth, as well as the current year, and yet markedly underestimated their ages and hospital stays. Only schizophrenics were affected. Dahl considered these failures to note the ravages of time as examples of the schizophrenics' "pre-Aristotleian paleologic," in contrast to "Aristotleian logic, the only one known to our society." However, he did not exclude the presence of a cognitive defect.

Working in Britain in the late 1970s, Timothy Crow and his group found that a full quarter of long-stay chronic schizophrenics perceived a five-year, or greater, discrepancy between their true and stated ages. He went on to much more complete mental status testing of these patients, and compared them with patients who were more accurate—those who were correct to within five years (over half of these patients were off by between one and five years). Using this very simple point of differentiation, the differences between the two groups were found to be rather startling. All were asked their year of birth, duration of stay, present year, age at admission, and year of admission. Over 60 percent of the not very well oriented "controls' got all five questions right; only a few percent of the more disoriented answered more than three of five correctly. Among them were a very few who knew their year of birth as well as the present year, yet markedly underestimated their age and stay. For these few patients, time may have subjectively "stood still" for reasons other than cognitive difficulty. For the large majority, this was not the case. Crow and his colleagues have continued to point out markers of this subgroup of patients, whose intellectual deficit may well represent organic impairment.[7]

Such a development is by no means unprecedented. It has been known for a long time that some chronic schizophrenics, particularly those with "negative symptoms," are marked by intellectual deficits. Such ideas go back to the nineteenth century and Hughlings Jackson. One common reading of twentieth century psychiatric research is an attempt, increasingly successful, to find "organic" dysfunction and pathology for what has been called functional illness. It is often forgotten that until rather recently epilepsy was a "mental" disorder, not a "neurological" one. It now appears that a sizable number of chronic schizophrenics are incapable of accurately reading a clock or a calendar. This is particularly significant, as an inability to read clock time probably affects the other time senses in a marked manner—whether internal time estimation or subjective time in general. It is part of general clinical practice that "organically impaired" individuals are often time disoriented. Patients with organic brain syndrome, despite its nature as an unspecific diagnosis, tend to overestimate time in studies of internal time estimation.[8]

If a proportion of schizophrenics appears to have features of organic disease, this itself will bias studies of time estimation applied to that population. In summary, we have to conclude that not all schizophrenics have similar deficits, at least in the chronic population. As the work of others has shown, an acutely presenting psychosis versus a treated psychosis, even over a matter of days, leads to very different results in tests of time estimation.[9] In looking at studies of time sense in schizophrenics, several questions should be asked: (1) What are the criteria of diagnosis? (2) Is the population chronic, and if so, are its members intellectually impaired? (3) If they are acutely ill, how acute are they; have they been treated, and to what level of change?

Internal Time Estimation

A hundred years ago, internal time estimation was the time sense most commonly studied, and it remains the most popular area of investigation. The most important American work on the methodology of internal time estimation was done in the late 1940s by Johs Clausen. Interestingly enough, his study subjects were chronic schizophrenics, pre- and -postlobotomy.[10]

Internal time estimation is a test to measure an individual's ability to note time duration without the benefit of an external clock.

Any duration of time can be treated, and researchers have considered periods varying from tenths of seconds to many weeks (one French researcher spent nine weeks in a cave, and estimated his stay when he came out as 36 days).[11] The actual methods used to estimate internal time are legion, often involving a series of comparisons. Luckily, however, they tend to break down into three major types:

1. The method of reproduction. With this method the experimenter creates a specific duration of time, using any methodology desired—a stopwatch, an electric buzzer, a lighted screen, or whatever. The subject is asked to reproduce the same time interval, often by tapping or counting. Over the past 20 years, the method of reproduction has come to be used less and less.

2. The method of production. By contrast with reproduction, the method of production has come to be used in a higher proportion of recent studies. With conditions set by the experimenter, the subject performs some action for a set period of time. Thus the experimenter will have the subject count, or tap, or take dictation, for what the subject thinks is the specified period of time. The critical factor is that the subject must produce the length of time, by him or herself, *de novo*.

3. The method of verbal estimation. Perhaps the simplest method for estimating time, the method of verbal estimation is very frequently used in time estimation studies, in an enormous series of variations. What is essentially involved is that some duration of time is placed before the subjects. They then estimate how much time has passed, verbally telling the examiner how long they thought the duration was.

Though in theory these tests appear simple enough, in practice they are not. Subjects may be grouped in a myriad of different ways whose forms must be specified. More significantly, the nature of the stimulus, the conditions of response, and the forms of intervention are almost infinitely variable.

Some facts do stand out. The three methods of time estimation are probably estimating very different things. For example, the method of reproduction simply does not correlate with the other methods.[12,13] The correlations that are obtained are not only nonsignificant, but most often nearly random. Why reproduction is so different from other forms of estimation can only be conjectured.

Whatever else is going on in the mind of the subject, reproducing an interval by tapping or counting may introduce a rhythmic element. Such an element may allow for greater accuracy and reproducibility of the method of reproduction compared with the other methods, which are themselves far more variable from test to test, that is, they have much greater intraindividual variation.

The other odd fact is that the methods of production and verbal estimation negatively correlate. Why this is so has been frequently debated,[14] and one consideration may have particular merit. While waiting for a stimulus to end, the method of verbal estimation may be prone to "influence" by a person's conception of subjective time. One can easily see that waiting for an intentionally boring tape to finish playing is probably a far more passive process than being asked to count out the same duration as on the tape. The variance and variability of verbal estimation are almost always greater than for other methods, and often enough its reliability may prove to be small.

Then there is the thorny issue of semantics. What is overestimation of time, and what is underestimation? One of the odd horrors of reading the literature on time research, often noted by investigators, is that one person's overestimation is another's underestimation, and vice versa. At one point Doob tried to clear up the confusion, only to produce a dizzying table that makes sense only with effort. The problem is a real one, and will not go away. I will follow a simple rule of reporting here—a subject's estimate will be compared with the objective interval. If the interval is 30 seconds and the subject says it is 40, that is overestimation; if the subject says it is only 20 seconds, that is underestimation. Some of the schemes that have been used to classify overestimation and underestimation of time rely on a belief that there is some kind of symmetry between "external" and "internal" clocks.[15] These schemes usually assume that if someone says a 30-second interval is 20 seconds, that person's internal clock is going "slowly." There is very little evidence that such global internal clocks, even if they do exist, mirror external time directly or indirectly; there is certainly evidence against such a model.[16]

There is one more preliminary consideration, the length of the intervals estimated. As with normals, most studies of mental patients are carried out for intervals varying from a few seconds to a couple of minutes. Even within such short periods, a few disclaimers have to be made. Vierodt's law, first proposed in 1868, *generally* seems to hold, in that people tend to overestimate periods of the order of five to ten seconds, while underestimating periods of a half minute or more. When the intervals studied are for ten minutes or longer, results become much more variable, though not less accurate, with the variability increasing as time length is increased. In general, for periods of 20 minutes or longer, underestimation is more common than overestimation.[17, 18] The variability of underestimation and overestimation is only compounded by the variety of means of testing that are used.

Some of the earliest studies of short-time estimation by schizophrenics appeared in the 1930s. This early work found that normals would judge five- and ten-second intervals rather accurately, while schizophrenics overestimated them. At intervals of 30 seconds, the opposite was true, with normals overestimating and schizophrenics underestimating. The population worked with was presumably that of chronic patients. Clausen, when he worked with schizophrenics, found them generally accurate when estimating intervals of five to 30 seconds. Lobotomy seems to have changed results very slightly, or not at all.[19] Though literally dozens of studies have followed since the 1950s, they have mostly shown that at short intervals schizophrenics estimate very similarly to normals, though with greater variance.

The only major exception to the finding that short-interval time estimation is relatively similar in schizophrenics and normals is that of Melges, an exception that must be considered because of the superior methodology of his work. The schizophrenics he studied underestimated 30 seconds, using the method of production, by more than half. However, this discrepancy may be more apparent than real. Melges' group, which was exhaustively and intelligently studied in several other ways, had been admitted in an acute psychotic state. In most cases they had been treated for a matter of only a few days. It may be that when acutely ill, schizophrenics, if they do indeed fit long-term criteria for diagnosis of schizophrenia, may act as other acutely psychotic patients and underestimate short intervals of time.[20] People who are acutely and floridly psychotic have a host of cognitive difficulties, and may only with great difficulty understand the procedures of testing.

As with normals, schizophrenics' ability to estimate time internally is affected by the conditions and interventions set by the tester. In one of the more unusual variants, while asked to estimate intervals from five to 30 seconds, schizophrenics were shown a series of drawings on a tachistoscope that were designed to upset them. For patients with "conflicts around aggression," a special design appeared of a "huge scowling man with manacled wrists and clenched fists raised high, ready to strike down upon an unsuspecting, smaller man standing in front and facing the other direction." Unsurprisingly, some patients' internal time estimates then became wildly off. On some level distress and pain impact on both objective and subjective measures of time.[21]

Studies of internal time estimation for periods of several minutes or longer also give results that show schizophrenics not markedly different from normals. On those occasions where overestimation was apparent, it occurred in populations that had been hospitalized for longer periods.[22,23] One study, subtyping schizophrenics into "paranoid" versus "withdrawn, thought-disordered" patients, showed marked discrepancies, with the "non-paranoids" more prone to underestimation.[24] What this finding means, without reference to age, criteria of diagnosis, length of diagnosis, and cognitive factors, is difficult to know. It may be that the overestimation of longer intervals sometimes found in "older" chronic schizophrenics may result from cognitive impairments.

Subjective Time Duration

Very few studies have considered the subjective time duration of schizophrenics in the nonacute state. Those few results that are available show that schizophrenics generally resemble normals with regard to the subjective duration of time, at least when they are not floridly psychotic.[25]

Subjective Time Perspective

There is a great deal of clinical evidence that subjective time perspective is painfully disordered in many schizophrenics, particularly in the acute stages of disease. What does occur in such stages of schizophrenia is very difficult to quantify, and is perhaps beyond the capacity of most testing procedures to describe. What is common during acute schizophrenia, unlike other illnesses (with the possible exception of affective disease), is the extreme sense of temporal disorganization that may be experienced. When it has been studied, this temporal disorganization tends to correlate with first-rank Schneiderian symptoms.[26] This finding seems to make sense—the more delusional and disorganized the psychosis, the more profoundly disorganized the sense of time. Whether one is causal or influences the other, or whether they both operate as phenomena with a common causal origin, is only conjecture at this point.

Quantitative studies can give little of the flavor of these temporal disorganizations. When time disorders and fragments, the world may become a uniquely terrifying place. It is an experience which, in its earliest stages, may be marked by an inability to relate subjective to objective time. The individual may see the hands of a watch "jump," much as one patient told me, "I turned around and did something and looked at my watch, and it just jumped an hour and a half." Or the process may take on far more onerous characteristics, as in this experience of one of Lehmann's patients:

> This is radio time—now 9:20." I hear this, knowing there is no time or day, but as I look at my watch—this thing on my wrist—the dial reads 9:20! But I cannot tell time—there is no time . . . "Time now 9:47 . . . time now 10:17." Every few moments or stretches of eternity this voice—the radio—says "this is radio time now" . . . It is real—it is not real—I do not know, but my watch seems to correspond exactly. Strange, strange, something must be bigger than I—some force greater . . . There are no days; no nights; sometimes it is darker than other times—that's all. It is never quite black, just dark grey. There is no such thing as time—there is only eternity.[27]

Subjective time is itself a strange occurrence, partaking of the uncanny. We generally inure ourselves to its strangeness by comparing it with "real," objective time, and studying our watches. When the "corrective" presence of clock time breaks down, the measure of time and what it signifies may evoke much disturbance, even terror. Though it is difficult to study acutely psychotic patients, clinical descriptions of the breakdown of the time sense in schizophrenics have a peculiar, specific ring to them. For these patients there is a succession of events where time telescopes and dives, moving reasonlessly from quick to slow, from timeless and eternal to stopped dead.

Whether the temporal disturbance of acute schizophrenia is very differentiable from affective disease is hard to know at this time. One may guess that temporal disorganization continues in some schizophrenics with chronic processes, acting as one element in the thought disorders they may possess. At least in its longer term forms, schizophrenics seem to have a time derangement rather different from those with affective disease. The time derangements of the latter move not haphazardly, but backward or forward into time. They are also marked by extreme derangements of subjective time duration, where one feels time as either very slow, as in depressives, or very fast, as in some, but by no means all, manics.

What has been extensively studied in schizophrenics, via a large series of psychological tests, has been what is generally called "future time perspective." In general, psychologists and psychiatrists have not tried to obtain quantitative measures on the subjective perspective of mentally ill patients. Despite their number, the psychological tests used have many methodologic drawbacks, which are serious whether they are used in psychiatric or normal populations. With the most commonly used of these projective tests, what they purport to measure may be different from what they actually measure.

One of the most common methods for describing "future time perspective" is that of "extension." Extension is supposedly how far one looks into the future, or back into the past. The test is more usually concerned with "future extension," and consists of giving subjects ten to 15 statements of "common life events" on a single card. The subjects are then asked to enter the age at which they expect to experience each of the "common life events." The furthest age forward listed by the subject is considered an image of "future time perspective," and is subtracted from the real age to give the person's "future time extension."

The same methods are then followed up for another common test, that of "future time coherence." Here the same ten or 15 "common life events" are each placed on separate cards. A few minutes or longer after filling in the answers to these questions, the patient is asked to arrange the cards in the order in which he or she expects to experience these "life events." Of course, the patient may well arrange them in a similar order to the one previously written down, ordered there by age. The correlation between the rank order by ages and the subject's later arrangement of the cards according to expectations, is a test of the subject's "future time coherence," and presumably tests the unity of his or her plans for the future.[28]

In tests done with schizophrenics, their "extension" is generally found to be less than that of normals. "Extension" is also less for those chronically hospitalized as opposed to those who have been hospitalized for only a short period. One study among age-paired schizophrenics and "normals" found the schizophrenics' "future time extension" as 35 years, versus the "future time extension" of normals of 49 years.[29] It is not very clear what this really means. Using the same methodology, an Israeli study compared chronically hospitalized schizophrenics and chronically hospitalized tubercular patients. "Future time extension" was 12.7 years for the schizophrenics, who had been hospitalized longer than the controls, whose "future time extension" was 15.5 years.

A very common variant of the extension test is to ask people to tell a story with structured elements, usually modeled on the Thematic Apperception Test (TAT). Here schizophrenics tend to give stories limited in future time to short periods; however, normals tend to give only slightly lengthier responses.[31]Another common test of extension is that of future event "density." People are asked to write down as many events as they can that will happen in the future. Groups are then compared, most often on the basis of how many events they have decided to write down.[32] Despite its wide use, this easy method is probably more a test of attention than a test of future time perspective.

The measurement of "coherence" may be more interesting. Normals tend to have high "coherence" scores, generally with correlations greater than 0.9. The results are very different for older, chroinc schizophrenics. Let us consider the plight of a chronically hospitalized patient. He is given a questionnaire, and reads such questions as, "How old might you be when:

"You retire?"

"Your last sexual intercourse occurs?"

"You can say that you have most of the things you want?"[33]

How will that patient react? You have someone hospitalized for a period of several years in a place where sexual activity is not allowed. The idea of retirement does not apply when you do not have a job. It may be difficult to answer such questions, or even to take them seriously. Coherence is really something different from "future time perspective," something that may have little to do with any feature of time perspective per se. What the procedure does do is test the ability of a patient to remember his or her responses to a questionnaire, and then to arrange these same answers in a different format. Presumably most "normals" recognize that the cards that they are given, with one question to a card, are listing the same questions they have seen before. The coherence test examines the ability of subjects to order events mentally and retain the memory over a few minutes. It does not measure the future time perspective of schizophrenics, or even how disordered that future sense might be, but their intellectual capacities to remember and to order. For some chronically hospitalized patients, particularly the subgroup that is intellectually impaired, that may prove a difficult task. For others, level of attention may prove the limiting factor.

Though projective tests of future time perspective give some data about schizophrenics' experience of time, they generally only show that it is different from normals. The clinical usefulness of such tests is doubtful. To obtain a flavor and an understanding of these differences, it is best to perform a clinical interview. Then one can note the enormous variation with which time is ordered and disordered in schizophrenic patients. Some subgroups of schizophrenics may possess some type of intellectual deficit that makes clock time difficult to read and comprehend, but most schizophrenics, when not acutely ill, seem to deal with clock time and internal time estimation in a manner similar to normals. Subjective time duration, again when not in the acute state, may also be rather similar. The real changes in schizophrenic time sense probably occur in subjective time perspective, both in acute and chronic illness.

The jumbling of time that schizophrenics experience almost invariably contains references to events that are considered symbolically important in their lives. This is the same as the order of subjective time perspective, where people define proximity not by the linear distance of an event in relation to absolute time, but by the event's personal emotional significance. Such an ordering may be most clear, as well as unbearable, during the acute process of the illness. An event that happened 30 years ago may feel suddenly as if it just happened, or that it will never happen, but resolve only at the end of eternity. This bizarre, intermittent fragmentation of time for schizophrenics is perhaps the most salient feature of their temporal disorganization. It may be that in acute schizophrenic psychosis, the sense of subjective time, and what is important in subjective time perspective, overwhelms all objective time considerations.

DEPRESSION

Compared with schizophrenics, relatively few studies of time sense have been performed with people suffering from depression or hypomania. Why this is so is not clear. A change in biological clocks, if not an outright cause, is felt by some to act as a major element of affective disease. Furthermore, in the early clinical literature, changes in time sense were reported both more widely and more prominently in depressed patients, as compared with schizophrenics. Yet the background of reported knowledge is far less extensive.

External Time Calculation

The ability of depressives to read clock time has rarely been studied. This is both surprising and distressing, in that pseudodementia is a common finding in cases of severe depression. Often in pseudodementia the ability to estimate time and to maintain spatio-temporal orientation is markedly impaired. In view of recent reports of positive Babinski signs in cases of severe depression,[34] we may begin to wonder whether more severe cases of depression occasionally involve deficits of cognitive function whose etiology is overtly organic, in ways we presently know very little about.

Internal Time Estimation

The only temporal area in which depressives have been studied with real thoroughness is that of internal time estimation. Over short intervals of time, 90 seconds or less, depressed patients have an ability to estimate internal time that is equal to that of normals. The differences between depressives and normals, where they exist, are less than between normals and schizophrenics.[35] At intervals beyond several minutes, there is some question as to whether depressives do in fact overestimate, for the data here are sparse and contradictory.[36,37] These different results may be attributable to the differences of technique in the way the tests were administered, the severity of patient pathology, and the general inhomogéneity of the patient populations.

Subjective Time Duration

Changes in the subjective time sense of depressive patients were first noted about 100 years ago, and have been a common clinical feature since that time. What is generally expressed is a sense of a remarkable slowing of time. Feelings of time stopping, or the present acquiring the sense of the eternal, may also occur, if perhaps less commonly than in schizophrenia.[38] The change in subjective time duration is so prominent for many depressives that it became the subject of one of the rare examples of psychiatric poetry about the experience of patients:

> The hours drop slowly one by one,
> Into a void of vacant days and nights
> That stretch things endless ways
> Between the rise and set of sun
> No sound returns from that abyss
> Where fleeting time has ceased to hum,
> For ages there is nothing done
> And never will be, I wis.[39]

The feeling of endless, slow time is an accurate picture of the temporal sense felt by many depressed patients. In the few quantitative studies that have been done, between 60 and 80 percent of depressed patients describe the world about them as moving slowly or very slowly, in a highly sustained manner.[40] What is even more interesting, these same patients retain an ability to estimate internal time accurately.[41]

Subjective Time Perspective

When projective tests were run with depressives, the results were very consistent. Though we may not trust the findings of studies using "extension" tests and other techniques, the uniformity of result is impressive. In general depressives do not look far forward into the future. It is as if the future is blocked, and does not matter.[42] On scales devised to test temporal organization, profoundly depressed patients do very poorly.[43] All these findings merely reinforce clinical experiences one has with depressed patients. As one patient put it:

> The future to me is remote. I feel hopeless. I could look for the future, but I can't now . . . I want to get something back to my mind that seems to have gone, to let me see the present and the future, rather than to keep me looking toward the past. There is in me a kind of routine which does not permit me to engage the future.[44]

The most powerful time changes seen in depression involve subjective time duration and subjective time perspective. With the exception of pseudodementia, objective time is little affected. "Now" appears to go ever more slowly, and the future is blocked off, incomplete, hopeless. It is pointless to think about the future; there is nothing one can do. What is perhaps most surprising about these distortions of subjective time sense is that measures of objective time, particularly internal time estimation, remain so intact. Here depressives appear similar to normals. It is as if they understand, and can objectively orient themselves to the speed of the world, retaining their capacity for objective time judgment. Yet the future is hopeless, they are worthless, and objective time is without meaning.

The ability to appreciate objective time correctly, while subjective time is experienced as infinitely slow, has been seen in a variety of disparate conditions, including after bilateral thalatomy.

This juxtaposition of objective time accuracy combined with subjective time distortion and slowing is very different from schizophrenia. With schizophrenics the two senses of objective time and subjective time may prove difficult, and at times harrowing, to reconcile. In severe depression it appears that objective time and subjective time continue, but are somehow disconnected. They both seem to exist, but do not appear to affect one another. Part of the reason for this seeming disconnection may be because of the basic clinical picture of depression, where disturbances of affect overwhelm the concerns and considerations of daily life.

MANIA

Relatively little has been done with depressed patients, but even far less has been done to study the time senses of manic and hypomanic patients. What results do exist suggest that manics appear opposite to depressives in some features, though the degree of difference from normal is less pronounced.

External Time Calculation

Manics who have been studied have generally been looked at during the hypomanic state. In such studies their ability to read clock time has not been examined.

Internal Time Estimation

In the mid-sixties, Mezey and Knight hypothesized that hypomanic patients possessed internal clocks that were "set faster" than those of normals. They thereby predicted that hypomanics would underestimate internal time, as compared with normals. Such a result did in fact turn out to be the case, but not quite as the investigators had expected.

All the subjects were asked to estimate 30 seconds by the method of production. The "normals" consisted of young, junior staff at the Maudsley clinic in London. When asked to estimate 30 seconds, these normals gave highly reliable, reproducible estimates of 40 seconds, an overestimate of a full ten seconds. Asked to perform the same task, the hypomanics estimated 29 seconds. The hypomanics were certainly faster, but also a good deal more accurate than the "normals." Perhaps more of note, the same procedures were used with these hypomanic patients after "recovery." Opposite to the hypothesized results, these "recovered" hypomanics, presumably "slower" than when they were ill, gave shorter estimates (about 26 seconds) on retesting. Their estimates for 30-minute intervals also tended to be more accurate than those of normals.[45]

Subjective Time Duration and Perspective

While manics, like depressives, clearly retain the ability to tell objective time whether well or ill, they also exhibit changes opposite to those in depressives in their sense of subjective time. When ill, a majority of manics report subjective time as passing fast or very fast, and when recovered, report it as slower.[46] When normals are asked to describe their sense of subjective time, they usually report it as "normal" or "fast." Such findings in adult normals, generally tending toward the "fast track" with regards to subjective time, thereby make the difference between normals and hypomanics less than the difference between normals and depressives. We may conjecture that manic behavior is more acceptable and when controlled perhaps even useful—at least in industrial societies. People generally prize energy, and the ability to work that it gives. As in depression, however, the major difference between manics and normals lies in their feelings of subjective time. When ill, many manics view the world as a fast place. Unfortunately, studies on temporal organization and subjective time perspective among manic patients are very rarely reported.

ANTISOCIAL PERSONALITY

A great deal has been written on the time sense of antisocial personalities, "maladjusted children," juvenile delinquents, and the like. This may seem odd to some clinicians, who have a great deal of difficulty deciding whether these categories constitute "diagnoses" or simply exist as "social phenomena." The issue of whether there is a clear diagnosis may be far less significant to nonclinicians; the problem is clearly present, and is eminently worth studying. Yet the difficulty of just what sort of population is being investigated remains, and the issues of homogeneity, validity, and generalizability become more pressing. For logistic reasons most of the studies have been carried out with children and adolescents. Often the latter are incarcerated, and it becomes difficult to know, with jail terms standing as a main variable differentiating cases and controls, just what kinds of individuals are being discussed.

External Time Calculation

One of the more interesting studies of socially "disadvantaged" children, many of whom later became wards of social agencies, took place as part of the Detroit Group Project of the 1940s. Through this program psychoanalytically oriented psychologists attempted to understand just what it was that caused some children undue hardship in getting through the early parts of life, particularly when others with the same background and social setting were able to succeed in doing so.

Some of the children, though about to become teenagers, just could not read a clock:

> Several of our children never acquired any concept of objective time, couldn't read a watch, confused yesterday with tomorrow, allowed weeks, days, months, and years to slide into another with the abandon usually limited to the very little child.[46]

This inability to learn objective time left these children in the world of subjective time. If they played a ball game, time was "short" when they had possession of the ball. Without possession, time seemed abnormally long. After what they *felt* was a long time, they would demand possession of the ball, though in fact it was before the agreed-upon length of clock time was up. The investigators felt these children had no concern for the future. To tell them something had to be done in the future constituted, for them, "outright refusal."[47]

What the authors are talking about may or may not represent a feature of "delinquency" or "maladjustment," yet is still important. It points out that understanding objective time, particularly being able to read a clock, is a very necessary kind of "literacy" in trying to understand the world. In the Detroit Group Project, an inability to read clock time led to faulty internal time estimation. Life must be very different for those who live by subjective time, compared with the large majority who follow the ubiquitous watches and clocks of our society. It is not surprising that behavioral results follow, at whatever age the discrepancy appears. The problem may be particularly acute with the mentally retarded, or perhaps also for the learning disabled, who may have great difficulty becoming proficient in either measure of objective time. To live in a world of clocks and objective time, unable cognitively to appreciate that world and its forms of measure, must cause immense frustration.

Internal Time Estimation

There are few measures of internal time estimation abilities among "delinquents" in the literature, but what exists does show some of the pitfalls of such research. In the 1960s a series of evaluations of males aged 17 to 18 were performed in Israel. Two groups were matched for age, ethnic background, socioeconomic status, and education. The only "difference" between the two groups was that one consisted of inductees in the Israeli army and the other of inmates of a prison for "young offenders." In the first series of studies, verbal estimates of five, 15, and 25 seconds were significantly underestimated by the "young offenders," often by a factor of at least half. A few years later, a nearly identical group, asked to estimate the same intervals according to the same protocols, produced highly overestimated answers, often double the real duration. By contrast, while the "delinquents" had under-estimated as compared with controls in the first study, they consistently overestimated by comparison in the latter study.[48] The author could not explain this marked variation—in fact, outright contradiction—in these two results. Perhaps the explanation lies not in the subjects or controls, but in the inherent variability of many tests of internal time estimation, of which verbal estimation may be the worst offender. Replicability is often highly suspect, as, in turn, are the inferences that can be gained from such studies.

Subjective Time Duration

Most studies of antisocial personalities avoid quantitive study of subjective time duration. Anecdotal reports throughout the literature indicate that time goes both "very fast" and "very slowly" for criminals and juvenile delinquents, depending on environmental conditions. There is very little that can be conclusively stated.

Subjective Time Perspective

Many studies of time sense among delinquents are of future time perspective, utilizing projective tests with their manifold difficulties. Considering the generalizations these studies have produced, their methodology is worth describing, particularly in relation to their findings.

One of the more unusual, and often-cited, studies of this genre took place among "lower class" elementary schoolchildren in Pittsburgh during the early 1960s. The question was, given a "controlled" experience, whether those who "stole" and those who did not steal possessed different senses of time.

The children, 120 of them in grades three though six, were introduced to a "hearing tester," who was in fact the experimenter. She acted the role of an "apparently flustered" woman. Her handbag would be overturned and its contents would fall out; she then would say to the children:

> Look at this mess. I spilled my purse and everything fell out. I have to go up to the kindergarten now and I don't have time to clean this up. Would you mind getting these things together? I *don't care* about the change. I don't know how much it is but I think it is only pennies. Most of it rolled away I guess. Just try to get the papers together. I'll be back in about five minutes. (my emphasis)

Forty-nine of the 120 children "stole." After the return of the handbag, each was tested for "future time perspective." They were given a set of words about time, generally words denoting durations from a second to a year, and asked to tell TAT stories.

The results "confirmed" the authors' belief that time senses among the children who did and did not "steal" would be different. The "nonstealers" seemed to prefer "tomorrow" to "yesterday," as compared with their brethren who "stole." The "stealers" also told TAT stories whose "extension" into the future was one hour less than that of the "nonstealers." Interestingly enough, how much a student "stole" (it ranged from one cent to 50 cents) did not correlate with any of the temporal measures.

What does this study mean? Its ethics are clearly faulty, but the curious part is the double message given to the children. They are told by the examiner, a "hearing tester" they have never seen before, "I don't care about the change." It probably rolled away, she says, but "I think it is only pennies." Yet the examiner very clearly does care about the change—she is doing a study of it, in fact. We really do not know how these children interpreted her words. We also do not know what was happening in the lives of these children, what their families, linguistic sophistication, and social mores were really like. It is interesting that the stealers chose "earlier" time words than the nonstealers, but the inferences to be drawn from this are unclear. The inference that was drawn, that there is "strong support . . . for a positive relation between stealing and limited temporal orientation," is not justified.[49] Whether a child prefers tomorrow to yesterday, or tells TAT stories that last an hour and a half rather than two and a half hours, does not constitute a very useful understanding of "limited temporal orientation."

Similar difficulties crop up in a study of French schoolchildren, performed by an American, which used a battery of projective "future time orientation" tests. Here the study was between "maladjusted" children, "maladjusted" adolescents, and their "normal" counterparts. Whether the students were in special schools for those who previously had experienced "personality or behavioral problems" constituted the diagnosis of "maladjustment." The problem of diagnosis is a very real one here. Were these children learning disabled? Were they hyperactive? Were they "delinquents" who had trouble with the law? Although the author claims to have shown differences among the groups, most of his test results are statistically nonsignificant. Where his tests turn out to be statistically significant is instructive: with the children, on the time span of their TAT stories: and with the adolescents, in reference to "density." Of interest, the maladjusted children came from a lower socioeconomic background that the "normal" children. The older children were all from middle-class backgrounds. As for the differences between adolescents, the "maladjusted" adolescents listed a median of four events they could imagine occurring in the future, whereas the "normals" listed six. The "density" test was probably a measure of the seriousness and attention the students gave the various testing procedures.[50]

Finally, the question of matching must be considered, particularly for what criteria groups are "matched." An extremely well-matched study, carried out in California, coupled high school students for age, sex, ethnic group, socioeconomic group, and I.Q. Supposedly the only variable of difference between the groups was "delinquency" versus "nondelinquency." The authors then proceeded to give these students their own "future events test," a questionnaire consisting of 36 items that might happen in the future. The respondents were asked to state when, if ever, these events might happen to them, and to give the corresponding age at which they personally expected such an event to occur. When all had been added up, the average expected age for the 36 events was 29 years for the "nondelinquents", and 27 years for the "delinquents." Thus, the authors state, the nondelinquents "confirm the hypothesis" and "show a more extended future time perspective."

Do they really? And of what does this future time perspective consist? We learn a great deal more by looking at those of the 36 questions where the statistical differences are greatest between delinquents and nondelinquents. Such differences were most marked when the students were asked to estimate the age at which they expected to: finish college; get drunk; be hospitalized; graduate from high school; go to jail; have a friend die; win lots of money; have their first child; get married; have a flashy apartment.

We can particularly well understand the huge statistical difference between groups on when they will go to jail: the variable "delinquency" is itself defined as being incarcerated at least once by the police.

Does this test examine "future time perspective"? Is the age when one "expects" to finish college an element of one's subjective reach into the future, or of one's social judgment? If anything, the test produced by these authors shows "delinquent" youth to possess perhaps a rather realistic appraisal of their future. They may indeed get drunk earlier, or go to jail earlier, than their "nondelinquent" brethren.[51]

In sum, conclusions as to whether "delinquents" or "nondelinquents" have differing future time perspectives are marred by the methodologic errors of the studies themselves. What may be involved is not a question of whether differences of subjective time perspective exist,[52] but whether these projective tests can measure them reliably. It is felt, for example, that in studies done in the United States, class is not a factor in future time perspective.[53] Such a conclusion may be reached if one looks at the results of projective tests. Yet these tests do not consider the main elements of subjective time perspective—what events in the past and present are considered personally important and meaningful to the individual, and how these are used in creating the person's images of the future. On tests of "future time extension," the age of expected retirement for middle-class versus working-class individuals may be statistically very similar, particularly as such retirement ages are legislated. The meanings of these events may be profoundly different. Someone who is unemployed or rarely employed may write down a standard age for "retirement," but its meaning for that person will be different than for someone with a steady, lifetime job with a continuing prospect of advancement. Projective tests do not help us here. Clinically these nuances of subjective time perspective can be noted, and are important when we try to understand how people see their past and future lives.

THE STRANGE CASE OF NORMALS

In the literature of time senses in so-called normals, one is impressed not merely with the number and variety of such studies, particularly those of internal time estimation, but by the methodological difficulties inherent in them. A few reviews of such literature, particularly that on internal time estimation, do exist; Doob's enclyclopedic work is the best.

External Time Calculation, or Clock Time

Rarely do studies of "normals" and their ability to estimate objective time include data on their ability to read clocks. Most investigators probably assume that, if subjects cannot read clock time, they cannot be described as "normal." It is just accepted that the subjects are "healthy" and cognitively skilled. To assure such normality, most of the studies have used as subjects American university undergraduates, generally in groups of between ten and 30.

Internal Time Estimation

What is most impressive about the voluminous literature on internal time-estimation abilities of normals is the enormous variety of parameters used. To simplify matters, at least three major variables are involved—the set conditions of the test, the forms of intervention made by the examiner during or after each test, and the methods of measurement used.

With regard to the conditions of testing, many of the conditions devised in these tests are informed by a century-old debate about "filled" and "unfilled" intervals. A "filled" interval is one in which the subject has something to watch or do; an "unfilled" interval is presumably without either. As researchers have pointed out for decades, there is a marked unknown in such research designs. The behavioral paradigms of many of these studies have caused almost no one actually to ask these subjects whether the activity during these tests is considered interesting or boring, fulfilling or unfulfilling. We have virtually no information about what is going on in these subjects' minds when performing these tests.

For example, many studies consider "unfilled" intervals to involve reading a book, watching a light bulb, or sitting before a blank screen. By contrast, a "filled" interval will involve reading aloud from a book, watching a flashing light bulb, or looking at a screen with something on it. What is clear is that whatever stimulus is utilized, however boring or uneventful, human activity does not exist in a vacuum. People will be thinking of something, although it may have nothing to do with the experiment. All conditions are, to some extent, "filled."

When measuring normals, the three methods of internal time estimation are used in a dazzling array of variations, often with the added fillip of various forms of comparison and cross-comparison. As for intervention, examiners have utilized methods varying from openly lying to their subjects about what they are told to do, to reinforcing them with a huge variety of methods. Patients have been given drugs or electroshock; told to think of something pleasant or something awful; and paid for good performances.

It may now be obvious that the possible variations and permutations used to study people and their internal time estimation abilities are immense, perhaps infinite. Much of the enormous variation involves conditions of the experiment. People have been asked to estimate time intervals while in their beds, in caves, masked underwater, in university laboratories, in flight, in locked, darkened rooms, blindfolded in cars, in huge rooms while alone, and in small cubicles crowded with other people who vary in age and sex. They have been shown items on tachistoscopes and on screens, with various settings; with different colors of light, different speeds of flashing, different times of duration; with hearable tones increasing or decreasing in frequency or intensity, manipulable or out of range; while performing long division or copying from dictation; while holding the palm of the thumb backward or forward; while placing letters of the alphabet upside down; while performing tests of mechanical ability; while doing intelligence tests; while lifting barbells; while peddling on an ergometer; while listening to music—rhythmic, arhythmic, melodic, or unmelodic; and while expecting electric shocks that do or do not come.[54]

Varieties of reinforcement have included not merely electroshock, but payment and promises of future advancement. Subjects have been selected along a dizzying array of psychological variables, ranging from anxiety, stress, and tension, however defined, to different forms of personality—extroverted, introverted, or whatever.

To illustrate the extraordinary variety of such studies, here are a few examples. One study, of Australian females of unknown age, grouped the women on the basis of religious belief. Three groups were devised: those of "strong" belief, those with neutral positions, and those who were clearly "nonbelievers." All groups were then shown, on a tachistoscope, 17 "religious" and 17 "control" words, each five times for 250 milliseconds. Unfortunately, we are not told what the words were. The investigator expected that the "religious" women would estimate the "religious" words as taking longer than would the "nonbelievers." He was right, but the "nonbelievers" also estimated all the other words as taking less time than did the "believers." The investigator suspected the "nonbelievers" might have generalized their much shorter responses, as a result of having first been shown the "religious" words.[55]

A perhaps more involved study, here engaging personality type, was performed on 200 male summer students at a large American university. They were given a questionnaire, and when they had finished filling it out, were asked how long the process had taken them. The questions on the questionnaire were devised by the author to test "anality." Students were scored on how they responded to such statements as:

I believe in being thrifty.

I like to act stubbornly.

I believe in striving for perfection.

The students gratified investigator expectations. "High anal scorers" more consistently overestimated the time they took to fill out the questionnaire than did the "less anal" participants. Unfortunately the actual scores, as well as the actual estimations, are not recorded.[56]

Another interesting test of internal time estimation studied the variable "danger." Here undergraduates were placed blindfolded on a moving platform. The platform was made to move either away from or toward a precipice. We are not told just how far the fall from the precipice would have been. In both circumstances, moving away from as well as toward the precipice, the students were asked to push a button for "five seconds." Pushing the button had two results—a recording was made of how long the button was pressed, and, depending on the phase of the experiment, the time spent pressing the button was the time during which the platform actually moved. The experimenters expected that the "time estimations" of the students would be shorter when pressing the button moved the platform toward the precipice. They were right. Indeed the students took their hands off the button more quickly when it moved them, blindfolded, toward the precipice.[57]

One of the most ingenious studies of internal time estimation looked at the old issue of the effect of body temperature. An attempt was made to see if reduced body temperature did in fact lead to "slower" estimates of time. The subjects used were scuba divers, both professional military divers and skilled amateurs. Subjects were matched for age, sex, and nationality. The cold water was in Wales, the warm water in Cyprus. Here the investigator suspected that the cold waters of Wales also produced the confounding variable of anxiety. Thus, when the divers were asked to dive in Cyprus, a one-ounce charge of gun cotton, with the fuse lit, was placed in their mouths. They were asked to deliver the charge to a wreck lying in 15 feet of water. If the charge did not go off in a few minutes, the diver was meant to go back and retrieve the charge. Whether retrieval was ever necessary is not stated. The effect of lighted gun cotton in the mouth, at least on time judgment, appears to have been very small.[58]

Many of these experiments have been attempts to study one basic principle—whether increasing the number of "events" during a temporal interval increases the subject's estimate of the duration of that interval.[59] The results of these hundreds of studies are generally in agreement with this statement of Friel and Lhamon:

> Increasing number of objective events may increase temporal judgments when the events are perceived as discrete entities, but will decrease the judgments when the result is tighter organization, or, as has been suggested, stronger cohesiveness.[60]

The catch lies in definitions of "tighter organization" and "stronger cohesiveness"; these terms can be, and are, variously defined.[61] They also presume we know a great deal more about what is going on in the conscious and unconscious brain than we actually do.

Or perhaps Shelley said it best 170 years ago: "If a mind be conscious of a hundred ideas during one minute, by the clock, and of two hundred, during another, the latter of these spaces would acutally occupy so much greater extent in the mind as two exceed one in quantity." Or we can perhaps sympathize with the individual described by Doob:

> Is he bored, is he awaiting the millenium, is he waiting for the camera to click, is he wondering whether the plane can take off before it reaches the end of the runway, is he testing the accuracy of his watch, or is he just a subject in a psychological experiment trying to earn a little extra pocket money?[62]

To study the internal time estimation of normal subjects is only possible if an enormous host of variables—biological, social, economic, psychological, and others—can be matched and controlled. What is important is how little these different factors seem to change things. Such a fact is fascinating given that variables of real importance—arousal, attention, and concentration; affect, mood, and personality; and levels of perception and intellectual function—are themselves so hard to control for.[64] The human ability to estimate internal time appears to be remarkably stable.

Subjective Time Duration and Perspective

Though studies exist regarding subjective time in normals, they are rare, and are generally confined to tests of subjective time perspective. Such tests are generally of a type difficult to interpret.[65] Given the enormous variety found in "normals," little can be said conclusively.

6

Time and Psychoanalysis

Time as a subject holds an unusual degree of interest for psychoanalysts. As a group, psychoanalysts have been far more concerned with the subject of subjective time than have others, and their attempts to theorize about it are among the most sophisticated. The reason for this interest is not difficult to discern. Freud himself was fascinated by the "virtual immortality of the id," as well as by the "timeless unconscious." He held that to understand these elements would be to obtain "an approach to the most profound discoveries."[1] With such a path before them, it is unsurprising that many others, including Fenichel, Bonaparte, Eissler, Pollock, Loewald, and Hartocollis, have sought to follow it. Also some of these writers have expressed a similar unhappiness to that Freud himself was said to feel about the subject, as late in his life he frustratedly admitted that he had not been able to understand time.[2] During Freud's life and after, psychoanalytic theory has changed, progressing from the topographical model to the structural model, to ego psychology, and now to object relations theory, as successive stages of psychoanalytic thought. With these changes have come new attempts to understand the subject of time.

Many writers have taken Freud's pronouncements on the subject only from the cryptic statements noted above, or from the inscrutable statement that the sense of time "originates in our inner perception of the passing of our own life."[3] This is unfortunate, as there is much more evidence of how Freud thought on the subject, though it is mostly secondhand. The informant is Marie Bonaparte, who published one of the first psychoanalytic papers devoted to time in 1940.

It is indeed unfortunate that this essay is relatively little read at this date, as it has a quality of elegance. Whether discussing animals or children, soldiers or lovers, there is a warmth in Bonaparte's writing that transcends the heavy superstructure around which she writes. Some present-day psychoanalytic writing on the meanings of immortality, particularly that of Lifton, may have been influenced by Bonaparte.

The specter of the Kantian position on time and space often presents itself to writers on time, and Freud and Piaget were no exceptions. In *Beyond the Pleasure Principle*, Freud wrote that "as a result of certain psychoanalytic discoveries, we are today in a position to embark on a discussion of the Kantian theorem that time and space are 'necessary forms of thought.'" It is important to note that Freud saw the unconscious as timeless, and thereby felt the Kantian dictum to be wrong.[4] But what does it mean for the unconscious to be "timeless"? Bonaparte's explanation is that the unconscious is timeless because it cannot perceive time. In fact, the unconscious is too primitive to have *any* concept at all. According to Bonaparte, the unconscious is so early and so primitive that whatever we mean by "thought," in our conscious definition, has absolutely no referent in the unconscious.

Something as sophisticated as time could hardly be an attitude obtained via the unconscious, whose structures, if they changed at all, changed very slowly indeed. The very timelessness of the unconscious argued against its importance in creating the human sense of time. Bonaparte showed her paper to Freud, who responded not merely with a discussion of how the human time sense occurs, but with an entire outline of how human perception operates:

> In a conversation which I had with him after he had read this paper, Freud confirmed that his views were potentially in agreement with those of Kant. The sense we have of the passing of time, he observed, originates in our inner perception of the passing of our own life. When consciousness awakens within us, we perceive this internal flow and then project it into the outside world.

The quality of such projections, it turns out, is critical; in the brain most of what we call perception starts out as projections of different elements of the internal mind:

> The perception of space, Freud went on, cannot be separated from that of time. How have we come to acquire it? To begin with, we must ask ourselves whether there is anything in the world which we can conceive of apart from space, nonspatially. One such thing does exist, namely, the mind or psyche. But this discovery must itself provide us with food for reflection. If the mind seems thus to lack the quality of space, perhaps it is by reason of a massive projection outwards of all its original spatial attributes. Psychoanalysis has in fact taught us that the psyche is composed of separate institutions which we are obliged to represent as existing in space. But why should it not be the other way round? When our conscious begins to establish itself, it would perceive these internal institutions, the reconstruction of which we owe entirely to depth psychology, as located in space.[5]

So space is created as an outward projection of mind. Elements of this mind, Freud explains later, "possess an anatomical substratum," probably describing the structural model of id, ego, and superego.

Yet other perceptions are no different. They also are "projections," be they touch, taste, smell, or hearing, though each differentially distributed betwen inner and outer worlds. Even vision has all of its perceptions so "projected." Thus Freud concluded, "May it not be the same . . . with our external perceptions of space and time, and would not this translation into psychoanalytical language of the old a priori judgments of Kant vindicate him approximately?" So Freud has resolved any conflicts between Kant and psychoanalysis over the question of time, with what will become interesting results.

Bonaparte goes on in her paper to describe some of the basic attributes of time, yet the important conclusions have already been reached. Time and space are very much like a priori judgments. They are, perhaps, in the realm of Kantian noumena—their real nature can never be known.

Their real nature can never be known because of the understanding, which Freud and Bonaparte express on several occasions, that the unconscious is "unknowable." Not only is the unconscious timeless, that is, unable to perceive time, or conceive of time, but it is also so slow moving as to have been first thought by Freud as not to be moving at all. And the metaphor given to describe what changes can take place is a description of Pompeii after the eruption of Vesuvius, over the many ensuing centuries. The unconscious is "wholly inaccessible to direct knowledge and [something] about which we can only make inferences."[6] These few items of indirect knowledge occur "when the unconscious exhales from its depths, like the vapor which ascended to the Pythian priestess at Delphi, sudden gusts from its timeless and spaceless world, which then proceed to confuse and disorganize in our dreams our perceptions of space and still more of time."[7]

To see the unconscious is a more obscure process than looking "through a glass darkly." What Freud has done is to explain that the unconscious looks very much like the Kantian phenomenon of the "thing in itself." It is ultimately unknowable; only the tracks of its emanations can be watched, and then only after unending disguise. Its movements, when they do exist, are enormously slow, glacial. And yet, particularly in early psychoanalytic theory, the unconscious was the sole determinant in our actions in life. As Pollock has written:

> In the early stages of psychoanalysis, with its emphasis on id psychology, there was a decided tendency to understand psychic life as wholly determined by our unconscious past; unconscious forces from the past explain the development and vicissitudes of life.[8]

Is it any surprise that the process of analysis may proceed in a fashion "terminable and interminable"? The static quality of the psychoanalytic process becomes theoretically comprehensible when we consider the qualities of the unconscious. Even to modern-day analysts, many of the first causes in our daily lives reside in the unconscious and id. Yet both are "unknowable." Is it difficult to understand that many years may be required to investigate what is "unknowable"? And the unconscious is not merely unknowable, but is also "timeless." Even with the introduction of ego psychology and object relations theory, and the development of the concept of "compromise formation"[9] in attempts to explain intrapsychic life, psychoanalysis faces large uncertainties. In looking at early psychoanalytic theory in reference to time, we seem to find a situation frozen in ice.

Most analysts intuitively understand the high speed of much of human life, intrapsychic and beyond. Yet the theoretical concepts of psychoanalysis may perhaps not bear this scrutiny, and may give of their own accord a better understanding of the sometimes "endless" quality of psychoanalytic intervention. Elements of slowness and immobility weave their way through the corpus of psychoanalytic theory. In *Civilization and Its Discontents*, Freud declares, quite abruptly, that while the body adapts to evolution, the mind is somehow exempt. The mind exists with its state of evolutionary movement unchanged. "The fact remains that only in the mind is such a preservation of the earlier stages alongside the final form possible"; moreover, "we are not in a position to present this phenomenon in pictorial terms."[10] What gives the mind its unchanging stance, a "force" to resist evolutionary movement?

With a circadian view of biological time, brain movements and activities are extraordinarily fast, regulated by a huge number of equally quick actors, buzzing and colliding with each other. Contrast this speed with the shifting compromises between id, ego, and superego that are said to explain the bases of intrapsychic life. Both views are very far from providing the total picture. Both views may be complementary. They are also, at least on the temporal scale, highly distinct.

The psychoanalytic understanding of subjective time, at least on the theoretical level, is far more sophisticated and complex than any biological theory, and, particularly for the early years of life, pretty much the only one available. In general analysts see objective time, rather like Piaget, as arising out of subjective time. For the infant, nothing but subjective time is said to exist. For the analyst what constitutes time in early life comes not from objective reality, but from subjective reality—the child's intrapsychic life. What is most consistent in analytic discussions of how the time sense grows in the child is the importance placed on the infant–mother relationship. Some writers have gone so far as to speak of "Mother Time."[11]

Some analytic writers accept the importance of innate biological rhythms after birth;[12] others see things very differently. For these others, the importance of the biological lies in how it becomes psychological. This transformation is said to unfold in the early mother–child interaction.

For the infant the first "cradle" of time experience is said to be the periodicity of frustration and gratification.[13] Piaget as well sees this cycle as fundamental to the developing infant's sense of time.[14] Spitz saw this experience as beginning with the first cry of life outside the womb, the first declaration, as it were, of unpleasure.[15] For analytic theory the infant first learns about time through the hunger–satiation cycle. Time is that breakpoint in movement from hunger to satiety, that sense of moving back and forth from one to the other. The mother is both the source of frustration and the source of gratification; she is the first teacher of time.

For analytic theory, movement from the oral to the anal stage is particularly important to the developing child's sense of time. Time, as it relates to control and psychopathology, suddenly becomes very significant. The relation between "anality" and time, first pointed out by Freud, has been described by a host of writers, including Abraham, Jones, Meerloo, and Fenichel. The timing of excretion is viewed as the major battleground between mother and infant, as the fundamental crux around which the infant starts to gain a sense of objective time. To control time is to control the environment.

In the anal stage, the transitional object takes on lasting importance in helping to develop what will eventually become the subjective time perspective of the child—a sense of past, present, and future. Here again the mother's role is critical. Mahler states that the infant's sense of time is based on the absence or presence of the mother. Now, the present, means "mother is here." The representation of her absence is the image of the "past." When she will return becomes the nidus for the concept of the future.[16]

The development of object constancy is also critical to the child's learning a sense of time. With further internalization of parental figures, the child's sense of time not only becomes more specific, but more consistent. The idea of duration can develop, as the child now has some vague idea that past, present, and future are somehow continuous with each other.

Concomitantly the child learns to think about time with language. Studies along Piagetian principles determined that the word "soon" is generally first heard around 18 months. "Today," both as a word and as a concept, follows quickly after. Tommorrow and yesterday are used starting at about 30 months.[17] Thus while object constancy becomes established, the linguistic basis of a continuous time sense is also beginning. Whether the infant learns underlying ideas of linguistic time based purely on spatio-velocity concepts, or the other way around, is still a controversial debate, particularly in the nonanalytic literature.

The analytic view of the time development of the child is only a theoretical view. Still, in many ways, it is the only plausible theory available for the early period. Piaget himself stayed away from experimentally studying time concepts in children until about age four to four and a half. He may have done so on the basis of methodological considerations, as many of the interesting questions he asked about time could not be tested with children until they had at least a rudimentary conception of words involving time. Whether he resisted for such reasons or not, the Achilles heel of analytic conceptions of the development of the child's time sense lies in similar considerations. It is difficult to study and understand the nonverbal child.[18]

Within all this theorizing, the central role of the mother–child relationship stands out. Humans are animals posessing an enormously long educational and maturational phase. It is unsurprising that concepts such as space and time, so central to perception and brain function, receive their earliest notions from the earliest caregivers, on whom the child is totally dependent. But one must not avoid looking at the active functioning of the infant itself, its capacity to learn from the environment, and eventually manipulate it. For a long time that environment will be strongly bound up with the care givers, whoever they might be. However, equally early on, that environment may sometimes engage only the child, acting alone in a nonparental physical world.

It is perhaps a tribute to psychoanalysis that someone who has considered the problem of time from a multitude of perspectives now writes about time with ideas generated primarily from psychoanalytic theory. Originally coming to time from problems in the realm of physics,[19] J. T. Fraser now takes his philosophical directions mainly from the psychoanalytic camp. He has recently attempted to subtype the more primitive forms of subjective time that humans possess, and has developed a hierarchical, ostensibly evolutionary structure to explain their use.[20]

Fraser takes a cue from Jakob von Uexkull's concept of the *umwelt*, a concept of human perceptual limits. The umwelt is defined by the fact that "an animal's receptors and effectors determine its world of possible stimuli and action." A person's world and experiences are limited by the limits of perception. Fraser builds on this concept to offer a hierarchical series of "umwelts", themselves allowing the use of only certain stage-limited concepts of time.

First, for the animal, there is the world of the "biotemporal" umwelt. In this umwelt, time is without direction, without past or future, without memory, and all that exists is a "creature present." For humans there are a series of similarly primitive time conceptions. In one such conception events are isolated and separate in time, hanging together only by the thinnest of threads. There is a still more primitive human conception where time is so blank that it nearly fails to exist, and somehow melds into space. Fraser then locates the unconscious as operating in what he calls the "eotemporal" umwelt. Here time is similar to the *nunc stans*, the "abiding present," described by classical philosophy. Finally there is the highest level, or "noetic" level of temporal consciousness, something akin to what we call objective time. He sees the human mind as alternately "dominated" by subjective or objective time senses, the input of one versus the input of the other determined by contrasting investments of "psychic energy."

Fraser continues on to describe how concepts such as timelessness, time endlessness, eternity, and temporal "stopping" are all different. But his classification of these various "umwelts" remains only a classification. Subjective time, as well as objective time, are "functions" of ego, id, and superego, and these multitudinous forces somehow align themselves, mechanism unknown, into the nexus of felt time. Within such a scheme, an individual's concept of subjective time is not merely unfailingly unique, but moves formlessly from moment to moment, in ways that cannot even be guessed. Any attempt to generalize, any attempt to relate one form of psychopathology to a particular subjective or objective time state, receives no aid here.

Other psychoanalytic writers are attempting to understand specific disease states, and certain mental apperceptions of time, as involving specific parts of the mental apparatus. Peter Hartocollis has written that "time appears to move slowly when instinctual wishes remain unfulfilled or become repressed; and time appears to move quickly when wishes become converted into, or replaced by, superego demands and expectations that one resents." He also concludes that time can be experienced in only two major ways—as duration and as perspective. Duration is defined by him as strongly related to drives and their potential gratifications. Perspective is seen as entirely an expression of self and object relations. He has written on the different time senses seen in borderline patients, and at length on how time is put together in dreams. About dreams he writes:

> When time is explicit the conflict is likely to refer to an early event or relationship; when time is disguised or implicit in the dream experience, it may signify a current conflict, one involving a contemporary relationship or an early relationship reactivated in the transference.[21]

Is it really that clear? Most analysts would probably not deny that drives are involved in people's sense of duration, or that self-object relations are involved in one's sense of time perspective. But is that all? Most analytic writers have engaged more complex models when trying to understand an individual's sense of time. With regards to depression, in which people often describe distortions of their subjective time sense, Hartocollis declares "these are people with an obsessive-compulsive character structure who labor under a very harsh, primitive superego." Is this really true for everyone, even the "functionally depressed"? What about the patient with cancer, or the alcoholic, or the withdrawing cocaine addict experiencing depressive symptoms and a slowed sense of time?

Psychoanalysis has some very sophisticated models for understanding an individual's sense of time. However, the generalizability of these methods is severely jeopardized by the analytic method itself, which generally draws conclusions from a universe of one, or perhaps a few, patients. This leaves some psychoanalytic thought on time relegated to the place of speculation, no matter how plausible it may be. Large groups of individuals, let alone homogeneous populations, do not appear available for study by analytic methods, both for logistic reasons and because of the conditions of the psychoanalytic outlook itself.

The seeming inability of psychoanalytic research to adapt itself to presently used concepts of research design, engaging controlled populations and verifiable, reliable measures of outcome, is unfortunate. The intensity and sophistication of analytic technique allows a relatively greater understanding of just how complicated the individual time sense really is. Other theoretical constructs, particularly from the biological camp, may sometimes deny the systematic complexity of the time sense.

Other caveats should be introduced with regard to psychoanalytic concepts of subjective and objective time. For the analyst the unconscious may be a timeless, virtually unchanging, entity. This is not true of the conscious human time sense. Piaget and Fraisse have shown the many movements and fluctuations of a child's time sense, of both objective and subjective time, through the early years of life, including the oedipal period. If this is true, and memory is the stuff on which psychoanalysis exists, it follows that different memories and different residues of memory will all be seen through the prism of different time perspectives. The four-year-old does not see time and its meaning as the five-year-old does, or the eight-year-old, or the ten-year-old. Such cognitive changes in time and what they mean must be related to individual memories of these periods, a consideration that psychoanalysis sometimes ignores. Not only does the four-year-old differ from the five-year-old in his or her subjective and objective time senses, but in the whole way he or she uses time to construct and integrate memory. Not only do our memories change with age, but so do our capacities for them, and our ways of using them.

Since the advent of object relations theory, psychoanalysis no longer looks exclusively to the past to explain present behavior.[22] Still, the future has been inadequately considered by psychoanalytic theory. Much of what people live by and for comes from their images of the future, and where they will be in it.[23] Along with demanding more consideration for the idea of time in psychotherapy in general, how it is used, conceptualized, and what it means, the importance of the future must be recognized. The unconscious may be timeless, and may have stilled itself after the earliest years of life. But very often in human behavior, it is not the past that pushes us forward, but the beckoning of what will come.

7
The Future of Time Research

The experimental past of psychological time research extends back to Mach and Fechner well over a century ago, and includes some of the greatest psychological researchers. If anything, its past is richer than its present. The variability of the time sense with culture and environment, its susceptibility to elements as disparate as ambient temperature and personality, has caused many to throw up their hands in despair and so not study such an "impossible" subject. It is not "impossible." The last 30 years have produced new knowledge about human biological clocks that should revise our ideas of health and disease. When a circadian view of health does succeed, the clinical significance of psychological time necessarily will increase. Regardless of its present, the future of psychological time research is promising.

With psychological time research, one may begin to consider the odd, uncanny phenomenon of human subjective time, where a minute may feel like an hour, an hour like a month or a second. Subjective time, influenced by our internal biological clocks, our perceptual apparatus, and our environment, is an area where advance is almost inevitable, as so little is now known. Although it is fundamental to our memories, to our visions of ourselves and our lives, we have almost entirely neglected its study in favor of objective time. Psychological research has been far more concerned with having people tap their fingers, twirl barbells, and estimate the movements of stopwatches than with inquiring into how people personally experience time.

Perhaps part of the reason why this is so is that the psychological sciences have felt the need consciously to model themselves on more positivistic, "scientific" natural sciences, such as physics. Most people see the time of physics in very clear terms—that of geophysical time. It follows that researchers of psychological time have been most interested in two areas: people's capacity to know geophysical time, and their ability to estimate time internally, without cues. We call this latter process internal time estimation. For every 50 studies of internal estimation, whether of German undergraduates in 1896 or American undergraduates in the 1980s, there is at most one study of subjective time.

With physics as the primary model, it is curious how little interest scientists have shown in how physics deals with time. Historically physics has not bothered much with the question of time. Throughout much of classical physics, time can be thrown into equations with positive or negative signs—the results do not change.[1] Totally unlike biology, classical physics pays scant attention to time. It is as close as we can come, perhaps, to a "timeless" science. For classical physics, the conditions of the universe were the same, are the same, and will be the same, as is required of postulates that attempt to attain the status of "natural law." The theological underpinnings of modern physics may now be falling as a result of the onslaught of the new, grand unified field theories. Yet they remain an ironic counterpoint to the absorption of psychological researchers with the "true" time of the world, that of physics.

The strong interrelationships between our sense of time and our sense of language also helps explain the relative neglect of subjective time. Scientific researchers, when they do think of time, automatically consider the geophysical time of clocks. This is not true of some other individuals. Writers and visual artists think of time as a much more variable entity, changing directly with their inward reviews of themselves.[2] Many members of the general population feel the same.

Though we are unaware of it most of our conscious lives, we really operate with at least three separate clocks. Most of us accept as "true" time the time of official clocks, the time of Greenwich and astronomical observatories, passed on to us by millions of mechanical timepieces. As we do not always have clocks or watches directly before our eyes, we come to rely on our second "clock": our ability to estimate time internally. Our remarkable capacity for internal time estimation probably involves social cues and overlearned responses, as well as circadian rhythms, our internal biological clocks. Finally we reach that other clock, which tells us how time affectively felt to us as it passed—was it fast, was it slow, or did it not have any meaning at all? This third clock, our clock of subjective time duration, we experience most deeply during periods of great pleasure or great pain. On ordinary days, we generally disregard it in favor of objective time. Different cultures have different senses and values of all three clocks. Different people have different senses and values of all three clocks. Our social communications with each other are ultimately influenced by these different senses of time.

Subjective time is an entity by which much of our internal psychology and internal perception may be explained, if we bother to work carefully enough describing its forms and types. A carefully thought-out methodology is necessary. Many studies of time look at subjects whose diagnostic classification is given without defining the rudiments of that classification. What does it mean when an author tells us that all the people in his study were "antisocial personalities," "depressed," or even suffering "derealizations," if he does not define what those words mean? To deal with the psychology of time is to deal with language. We must use language precisely to describe what we are studying and thinking about. In the psychological sciences, one cannot rely on mathematically "clear" distinctions between phenomena, as one supposedly can in the physical sciences.

To use the scientific method, and our common statistical tests, we must work with homogeneous populations. To gain the advantages of experimental methods, individuals studied must be as alike as possible, with the exception of the one or more variables in which we are interested. In standard biological laboratory experimentation, where one can look at rice or fruit flies or peas that have been bred through generations for genetic similarity, we can usually assume such homogeneity. Human study is infinitely more complicated.

Unfortunately many researchers avoid the enormous difficulties of trying to make study populations homogeneous, and act as if no confounding variables exist. Thus many studies of time have looked at, say, a population of hospitalized people who vary in age from 19 to 82 years, and of whom 68 percent are female and 32 percent are male. Upon reading the results of this study, we find they refer to a person who is 42.6 years old, and who is 68 percent female and 32 percent male. No such being exists, but that is the person to whom our results refer. Such results would cause a problem if we knew that neither age nor sex had any independent or synergistic effect on the variables in which we are interested. But, we usually cannot assume that age and sex are unimportant; indeed, with regard to time studies, they are important. Different age groups have different time senses, and different abilities to keep their biological clocks in line. The problems in pursuing this research, in attempting to obtain populations as homogeneous as possible, are large. Yet when the confounding variables are known, one sometimes can correct for them, in ways that will not make selecting future populations hopelessly difficult.

Objective time is very different from subjective time; clock time is different from internal time estimation; subjective time duration, the sense of the velocity of time, is different from subjective time perspective, the way in which we place and organize our personal past, present, and future. Corollary findings flow from these more primary understandings:

1. Objective time is remarkably well conserved in humans. The ability of people to read clocks or to estimated time internally, is consistently retained, with the possible exceptions of individuals suffering from acute psychosis or severe organic lesions. There are brain-damaged people who feel the universe as hopelessly slow, moving before them in slow motion, who yet can still estimate internal time very well. Normal people placed in sensory isolation, made to go for days without sleep or swung around in Baranyi chairs, still possess an exceptional ability to estimate internal time. Why this is so is not understood. The overlearning of clocks in our society may provide a partial explanation. However, this hypothesis does not explain why individuals from cultures without clocks, and without even a word for time, can estimate internal time as well as Westerners can (see Chapter 4). We should look instead to our circadian rhythms, our internal, unconscious biological clocks, to help explain our ability to estimate internal time so consistently.

2. Different psychiatric illnesses appear to have characteristic "lesions" of subjective time. Generalized time slowing is particularly characteristic of major depression. When acutely psychotic, schizophrenics reportedly disassociate objective from subjective time, forming a clinical picture very different from depression. The clinical evidence available suggests that subjective time is very different in affective, as opposed to schizophrenic, illness. These findings demand an attempt at corroboration.

3. Emotion and affect can markedly change an individual's senses of both objective and subjective time. Expressions of objective and subjective time can be shifted by arousal and attention, by different levels of cognition, by age, by space, by distance, and by an enormous range of other environmental stimuli and conditions.

Then, we may ask, just where does subjective time come from? Is it a vestigial, prelinguistic element in our evolutionary growth of consciousness? Is it something hopelessly entangled in presently unnameable unconscious elements? How is it affected by changes in arousal and attention? Many theories of subjective time see it as most responsive to variations of pleasure and pain, broadly defined. When we are enjoying ourselves, time goes quickly. When we hate what we do, time goes slowly. People have different concepts of subjective time as they mature—what is the residue of these former time senses?

4. All cultures appear to possess objective and subjective time. Very little is known about the cultural differences of subjective time. With regard to objective time, the major difference from one culture to another appears in the precision with which time is measured. The capacity for internal time estimation seems to be universal. This universality is at least a first argument for the primary importance of biological clocks, as opposed to social cues, in the human ability to estimate internal time.

5. Humans probably start out with a sense of subjective time, and then go on to learn objective time. Both time senses change slowly and appear to mature over a lifetime. Followers of Piaget and Freud see objective time learning as a process occurring in definable stages. Learning objective time is affected by an enormous series of variables, of which space and velocity have been the most studied. Piaget's view, that humans learn objective time solely through the perception and abstraction of the velocity of objects, is at best a partial explanation. It certainly underrates the importance of early language development.

6. Like all eukaryotes, humans are circadian animals, but unlike most animals, we transcend our circadian clocks. Most of the organisms of the planet have used the daylight cycle as the focus of their internal time capacities. Circadian organization for virtually every biological task is retained throughout evolution. But humans are different—rather than controlled by their circadian clocks, like most of nature, they override them. We are time-using animals. The whole 24 hours of the day are our territory.

The key point in evolution is to persist. As Waddington and his successors have shown,[3] the capacity to persist is markedly improved by the flexibility of the organism, its capacity to live in different environments. Humans do not merely live in different environments, they create them, feeling and molding the earth simultaneously. They use time itself as an ecological niche. As industrial civilizations have increased in size, so have the inroads on our biological clocks, the results of which, in shift work and international travel, we are only now beginning to appreciate. Seeing human beings as circadian animals has profound implications for our practice of medicine and our concept of health.

A major goal of future time research should be to relate our unconscious internal biological clocks to our conscious clocks, what we call objective and subjective time. With such research the findings of physiology and molecular biology may someday link to areas as presently distant as personality, temperament, and affect. We are, of course, still very far from achieving such a goal. Relating biological clocks to our conscious expressions of time is an area where studies of people, both ill and well, normal and abnormal, will support one another, and where investigations of what first appear as arcane depictions of internal psychological states may directly relate to fundamental biological mechanisms.

To see if biological clocks do involve senses of internal time estimation, we can work in two directions: either studying individuals whose senses of internal time estimation are deranged, or investigating people whose biological clocks have somehow changed. The difficulty with the first approach is that most people whose sense of internal time estimation is inaccurate are themselves usually either brain damaged or acutely psychotic. At the very least, they tend to have some kind of larger cognitive or behavioral impairment. Nonetheless it may be possible to find some individuals with pure deficits of internal time estimation.

Subjects whose biological clocks are dysfunctional are more available for study. Individuals who suffer from what Aschoff calls "reentrainment by partition," usually people in experimental conditions whose activity–rest cycle has totally disassociated from their temperature cycle (see Chapter 2), might be one group to consider. Another group whose biological clock dysfunctions are less homogeneous, but still feasible to study, are travelers across multiple time zones. Also, bipolar manic-depressive patients, investigated for several biological variables at N.I.M.H., might be useful to examine for their internal time estimation ability. Through such studies we may be in a position to obtain leads as to where objective time originates.

To study subjective time is to investigate an area that is empirically almost blank. Given that emotion, affect, and arousal all probably help create our subjective time senses, it is probably useful to take a two-pronged attack on our capacity for subjective time by studying it in both normals and "abnormals" (the mentally ill). With normals it will be necessary to obtain baseline data on just how often, and how much, our subjective time sense varies during the day. How much does it vary from minute to minute, from activity to activity? Do people have the same general sense of subjective time from hour to hour, from month to month? Do individual differences show standard deviations similar to populations?

Beyond such "natural" experiments, we should also attend to the laboratory. Of particular importance will be work regarding boredom–interest, and the difficult subject of changing mood and affect. DeLong has already begun studies on the changes of subjective time brought about by differing spatial environments.

With regard to patients, an understanding of normals will undeniably help us to decide whether specific psychiatric syndromes show characteristic changes in the subjective time sense. First it must be seen whether the differences of subjective time sense noted in the literature really do exist. Are schizophrenics really so different in their subjective time duration from depressives? Are these changes manifested only during acute states, or when people are stable, but still profoundly ill? If these changes are not different from one disease syndrome to another, are they similar to situations sometimes seen in normals? Does "dejection" in normals do the same thing to the time sense as "depression" does to the mentally ill? Does waiting for convalescence after trauma surgery produce similar changes in the same sense as "functional" depression does? What about the time senses of the learning disabled and the mentally retarded? Are they "fixed" at an early, childlike state of development? Are the "mentally ill" similar to how "normals" are "at times"?

Another area for research is the learning of subjective time, both in early and later years. The impact of circadian rhythms on subjective time has not yet been considered. The results of studies of circadian dysrhythmias and objective time might then provide clues for research on subjective time.

Whether looking at normals or those afflicted with psychiatric disease, the need for simple, valid scales is crucial. Work in this area has been hampered by a series of global scales that combine elements of both subjective and objective time, giving quantitative results that effectively apply to neither. In looking at subjective time duration, one will have to start simply, considering the felt "velocity" of time. Then it may be possible to tackle the difficult subject of "timelessness," the feeling people describe in experiences varying from mystic states to winning footraces. With greater knowledge the even more difficult question of subjective time perspective may become possible to study. Engaging a host of variables, some of which are verifiable, many of which are not, subjective time perspective may start to be examined with the simple line experiments of Cohen (see Chapter 3), while recognizing how approximate knowledge of such phenomena will be.

In reviewing the clinical implications of time studies, it is best to separate short-term from far-term ones. If the subjective time senses of affective disease and schizophrenia are different, this may be a useful point of differentiation between these two important, yet relatively ill-defined syndromes. Different time senses may have more immediate uses in psychotherapy. Certainly asking patients about their subjective sense of time can be a useful aid to memory (and what is important in their memories). We can learn quickly how patients internally align and integrate their past, present, and future. From such a life "scenario," we can discover a great deal about how people cognitively operate: with what precision, with what logic, and with what fears. Melges has described very well the importance of thinking about time in social communication: the different time perspectives of individuals can easily cause them to fall into conflict.[4] From asking people about their internal representations of time, we may learn much about what to do therapeutically. We may need to reconcile some patients to an expected future; for others, there may be the need to reconcile them with their past. Teaching patients to look at their lives in relation to time and to use time in different ways, both cognitively and behaviorally, may prove very useful.

The long-term clinical applications of time studies to psychology are mainly in theoretical directions. The time sense is fundamental to any human cognitive system. We may learn a great deal by contrasting how the sciences, particularly physics and biology, differentially use time.[5] And through the study of biological clocks, we may learn relatively soon how time is organized on the molecular level. When we do, we will immediately know basic facts that are of much significance in determining how we perceive and act upon the world. From such knowledge of time, both in its minutest biological organizations and in its more extended psychological manifestations, we may attain some real understanding of just how we think.

References

PREFACE

1. Fraisse, P. *The Psychology of Time*. New York: Harper and Row, 1963.

2. Melges, F. T. *Time: The Inner Future*. New York: Wiley-Interscience, 1982.

3. Campbell, A., Baldessarini, R. J. Circadian changes in behavioral effects of haloperidol in rats. *Psychopharmacology* **27**:150–155, 1982.

4. Toulmin, S., Goodfield, J. *The Discovery of Time*. London: Hutchinson, 1965.

5. Piaget, J. *The Child's Conception of Time*. New York: Basic Books, 1970.

6. Bonaparte, M. Time and the unconscious. *International Journal of Psychoanalysis* **21**:427–468, 1940.

7. Cohen, J. *Psychological Time in Health and Disease*. Springfield, Ill.: Charles C. Thomas, 1967.

8. Doob, L. W. *The Patterns of Time*. New Haven, Conn.: Yale University Press, 1971.

9. Bergson, H. *Duration and Simultaneity*. Indianapolis: Bobbs-Merrill, 1965.

10. Fraisse, *op. cit.*

11. Lewis, A. The experience of time in mental disorder. *Proceedings of the Royal Society of Medicine* **25**:611–620, 1931.

12. Schilder, P. Psychopathology of time. *Journal of Nervous and Mental Diseases* **83**:530–546, 1936.

13. Fenichel, O. *The Psychoanalytic Theory of Neurosis*. New York: Norton, 1945.

14. Meerloo, J. A. *The Two Faces of Man*. New York: International University Press, 1954.

15. Brock, T., Del Guidice, C. Stealing and temporal orientation. *Journal of Abnormal and Social Psychology* **66**:91–94, 1963.

16. Clausen, J. An evaluation of experimental methods of time judgement. *Journal of Experimental Psychology* **40**:756–761, 1950.

17. Goldstone, S., Lhamon, W. T., Nurnberg, H. G. Effects of alcohol on temporal information processing. *Perceptual and Motor Skills* **46**:1310, 1978.

18. Melges, *op. cit.*

19. Doob, *op. cit.*

20. Lehmann, H. Time and psychopathology. *Annals of the New York Academy of Sciences* **138**:789–821, 1967.

CHAPTER 1

1. Cohen, J. Psychological time. *Scientific American* **211**:116–124, 1965.

2. Meerloo, J. A. *Along the Fourth Dimension*. New York: John Day, 1970.

3. Lehmann, H. Time and psychopathology. *Annals of the New York Academy of Sciences* **138**:789–821, 1967.

4. Woolf, V. *Orlando*. London: Hogarth Press, 1928.

5. St. Augustine. *The Confessions*. New York: Collier, 1961, p. 205.

6. Eddington, A. S. *The Nature of the Physical World*. New York: Macmillan, 1929.

7. Landes, D. S., *Revolution in Time*. Cambridge, Mass.: Harvard University Press, 1983.

8. Prigogine, I. Physics and metaphysics. *Advances in Medicine and Biophysics* **16**:241–265, 1977.

9. Toulmin, S., Goodfield, J. *The Discovery of Time*. London: Hutchinson, 1965.

10. Crow, T., Stevens, M. Age disorientation in chronic schizophrenia. *British Journal of Psychiatry* **133**:137–143, 1978.

11. See preface.

12. Melges, F. T. *Time: The Inner Future*. New York: Wiley-Interscience, 1982.

13. Cohen, J. *Psychological Time in Health and Disease*. Springfield, Ill.: Charles C. Thomas, 1967.

14. Cohen, J. Subjective time, in *The Voices of Time*. Edited by J. T. Fraser. New York: Braziller, 1966.

15. Coleridge, quoted in J. Cohen, *op. cit.* 1965.

16. Orme, J. Personality, time estimation, and time experience. *Acta Psychologica* **22**:430–440, 1964.

17. Mezey, A. G., Cohen, S. I. The effect of depressive illness on time judgement and time experience. *Journal of Neurology, Neurosurgery, and Psychiatry* **24**:269–270, 1961.

18. Cappon, D., Banks, R. Experiments in time perception. *Canadian Psychiatric Association Journal* **9**:396–410, 1962.

19. Spiegel, E., et al. Thalamic chronotaraxis. *American Journal of Psychiatry* **113**(2):97–105, 1956.

20. Lehmann, *op. cit.*

21. Efron, R. Temporal perception, aphasia, and déjà vu. *Brain* **86**:403–424, 1963.

22. Luria, A. I. *Higher Cortical Functions in Man.* New York: Basic Books, 1980.

23. Janet, P. *L'evolution de la memoire et de la notion de temps.* Paris: Chahine, 1928.

24. Fraisse, P. *The Psychology of Time.* New York: Harper and Row, 1963.

25. Piaget, J. *The Child's Conception of Time.* New York: Basic Books, 1970.

26. Cohen, *op. cit.* 1965.

27. MacNeil, B., Pauker, S. G. Incorporation of patient values in medical decision making, in *Critical Issues in Medical Technology.* Edited by B. MacNeil and E. Cravalho. Boston: Auburn House, 1982.

28. LeShan, L. Time orientation and social class. *Journal of Abnormal and Social Psychology* **47**:589–592, 1952.

CHAPTER 2

1. Aschoff, J. Circadian rhythms in man. *Science* **148**:1427–1432, 1965.

2. Minkowski, E. *Lived Time.* Evanston, Ill.: Northwestern University Press, 1970.

3. Aschoff, J., Ed. *Handbook of Behaviorial Neurobiology*, Vol. 4—*Biological Rhythms.* New York: Plenum, 1981, p. vii.

4. Halberg, F. Chronobiology. *Annual Review of Physiology* **31**:675–725, 1969.

5. Aschoff, *op. cit.* 1981.

6. Edwards, L. N. Persistent circadian rhythm of cell division in Euglena: Some theoretical considerations and the problem of intercellular communication, in *Biochronometry.* Edited by M. Menaker. Washington: N.A.S., 1971, pp. 594–611.

7. Bruce, V. G. Cell division rhythms and the circadian clock, in *Circadian Clocks.* Edited by J. Aschoff. Amsterdam: North Holland, 1965, pp. 125–138.

8. Pittendrigh, C. Circadian systems, in Aschoff, *op. cit.* 1981, p. 61.

9. Daan, S., Aschoff, J. Short term rhythms in activity, in *ibid.*, pp. 491–498.

10. Halberg, F., et al. Spectral resolution of low frequency, small amplitude rhythms in excreted 17-ketosteroids. *Acta Endocrinologica* **103**:1-54, 1965.

11. Aschoff, J. Annual rhythms in man, in Aschoff, *op. cit.* 1981, pp. 475-487.

12. Rosenthal, N. E., et al. Seasonal cycling in a bipolar patient. *Psychiatric Research* **8**:25-31, 1983.

13. Campbell, A., Baldessarini, R. J. Circadian changes in behavioral effects of haloperidol in rats. *Psychopharmacology* **27**:150-155, 1982.

14. Pauley, A. E. An introduction to chronobiology, in *Biological Rhythms in Structure and Function* Edited by H. Mayersbach, L. E. Scheving, and A. E. Pauley, New York: Alan R. Liss, 1981, pp. 1-21.

15. Sauerbier, I. Circadian systems and the teratogenicity of cytostatic drugs. *Ibid.*, pp. 143-150.

16. Takahashi, J. S., Zatz, M. Regulation of circadian rhythmicity. *Science* **217**:1104-1111, 1982.

17. Fuller, C. A., et al. Circadian rhythm of body temperature persists after SCN lesions in the squirrel monkey. *American Journal of Physiology* **241**:R385-391, 1981.

18. Reppert, S. M., Perlow, M. J., et al. Effects of damage to the suprachiasmatic area of the anterior hypothalamus on the daily melatonin and cortisol rhythms in the rhesus monkey. *Journal of Neuroscience* **1**:1414-1425, 1981.

19. Moore-Ede, M. C., Czeisler, C. A., Richardson, G. S. Circadian time-keeping in health and disease: Part 2. Clinical implications of circadian rhythmicity. *New England Journal of Medicine* **309**:530-535, 1983.

20. Aghajanian, G. K., Bloom, F. E., Sheard, M. E. Electron microscopy of degeneration within the serotonin pathway of rat brain. *Brain Research* **13**:266-273, 1969.

21. Aschoff, J. Freerunning and entrained circadian rhythms, in Aschoff, *op. cit.* 1981, pp. 81-94.

22. Pauley, *op. cit.*

23. Pittendrigh, C., Daan, S. A functional analysis of circadian pacemakers in noctural rodents. IV. Entrainment: Pacemaker as clock. *Journal of Comparative Physiology* **106**:291-331, 1976.

24. Aschoff, J., et al. Desynchronization of human circadian rhythms. *Japanese Journal of Physiology* **17**:450-457, 1967.

25. Czeisler, C. A., Moore-Ede, M. C., et al. Human sleep: Its duration and organization depend on its circadian phase. *Science* **210**:1264-1267, 1980.

26. Moore-Ede, M. C., Shulman, F. Internal temporal order, in Aschoff, *op. cit.* 1981, pp. 215-241.

27. Moore-Ede, M. C., Fuller, C. A., et al. Thermoregulation is impaired in an environment without circadian time cues. *Science* **199**:794-796, 1978.

28. Czeisler, C. A., Moore-Ede, M. C., Coleman, R. M. Rotating shift work schedules that disrupt sleep are improved by applying circadian principles. *Science* **217**:460–463, 1982.

29. *Ibid.*

30. Jouhar, P., Weller, M. Psychiatric morbidity and time zone changes. *British Journal of Psychiatry* **132**:231–235, 1978.

31. Colquhoun, P. Rhythms in performance, in Aschoff, *op. cit.* 1981, pp. 333–348.

32. Aschoff, J. et al. The influence of sleep interruption and of sleep deprivation on circadian rhythms in human performance, in *Aspects of Human Efficiency: Diurnal Rhythm and Loss of Sleep*. Edited by P. Colquhoun. London: English Universities Press, 1970.

33. Blake, M. J. F. Time of day effects on performance in a range of tasks. *Psychonomic Science* **9**:349–350, 1967.

34. Folkard, S. Time of day effects in school children's immediate and delayed recall of meaningful material. *British Journal of Psychology* **68**:45–50, 1977.

35. Monk, T. H., Folkard, S. Concealed inefficiency of late night study. *Nature* **273**:296–297, 1978.

36. Blake, M. J. F. Temperament and time of day, in *Biological Rhythms and Human Performance*. Edited by P. Colquhoun. London: Academic Press, 1971.

37. Eysenck, H. J. Personality and the estimation of time. *Perceptual and Motor Skills* **9**:405–406, 1959.

38. Hughes, D. G., Folkard, S. Adaptation to an 8-hour shift in living routine by members of a socially isolate community. *Nature* **264**:432–434, 1976.

39. Stephens, G., McGaugh, J. L., Alpern, H. P. Periodicity and memory in mice. *Psychonomic Science* **8**:201–202, 1967.

40. Rusak, B., Zucker, I. Biological rhythms and animal behavior. *Annual Review of Psychology* **26**:137–171, 1976.

41. Wehr, T. A., Muscettola, G., Goodwin, F. K., et al. Urinary 3-methoxy 4-hydroxy phenylglycol circadian rhythm: Early timing in manic depressives compared with normal subjects. *Archives of General Psychiatry* **37**:257–263, 1980.

42. Schilgren, B. C., Tolle, R. Partial sleep deprivation as therapy for depression. *Archives of General Psychiatry* **37**:264–271, 1980.

43. Wehr, T. A., Goodwin, F. K., Eds. *Biological Rhythms in Psychiatry*. Pacific Grove, Calif.: Boxwood Press, 1983.

44. Wirz-Justice, A., Wehr, T. A., Goodwin, F. K., et al.: Antidepressant drugs slow circadian rhythms in behavior and brain neurotransmitter receptors. *Psychopharmacology Bulletin* **16**:45–47, 1980.

45. Knapp, S. Ordering the fluctuation of tryptophan hydroxylase: Lithium and chlorimipramine. *Psychopharmacology Bulletin* **18**:39–43, 1980.

46. Rosenthal, N. E., et al. Seasonal affective disorder. *Archives of General Psychiatry* **41**:72–84, 1984.

47. McEachron, D. L., et al. Lithium and biological rhythms of serum calcium, magnesium, and PTH. *Psychiatric Research* **7**:121–131, 1982.

48. Knapp, *op. cit.*

49. Medina, J. Cyclic migraine: A disorder responsive to lithium carbonate. *Psychosomatics* **23**:625–628, 1982.

50. Weiner, H. The prospects for psychosomatic medicine. *Psychosomatic Medicine* **44**:491–517, 1982.

51. Schoener, T. W. Resource partitioning in ecological communities. *Science* **185**:27–39, 1974.

52. Goodwin, B. C. *Analytical Physiology of Cells and Developing Organisms*. New York: Academic Press, 1976.

53. MacLean, P. *A Triune Concept of the Brain and Behavior*. Toronto: University of Toronto Press, 1973.

54. Bonaparte, M. Time and the unconscious. *International Journal of Psychoanalysis* **21**:427–468, 1940.

55. Melges, F. T. *Time: The Inner Future*. New York: Wiley-Interscience, 1982.

56. Doob, L. W. *The Patterns of Time*. New Haven, Conn.: Yale University Press, 1971.

57. Melges, *op. cit.*

58. Lord Brain. Some reflections on brain and mind. *Brain* **86**: 381–402, 1963.

CHAPTER 3

1. Piaget, J. *The Child's Conception of Time*. New York: Basic Books, 1970, p. ix.

2. Piaget, J. *Insight and Illusions of Philosophy*. New York: The World Publishing Company, 1971, p. 57.

3. Piaget, *The Child's Conception of Time*, p. 274.

4. *Ibid*., p. 257.

5. Piaget, J. *The Child's Conception of Space*. New York: Humanities Press, 1956, p. 462.

6. Piaget, J. Time perception in children, in *The Voices of Time*. Edited by J. T. Fraser. New York: Braziller, 1966, p. 208.

7. Piaget, *The Child's Conception of Time*, p. 267.

8. Piaget, Time perception in children, p. 209.

9. Piaget, *The Child's Conception of Time*, p. 271.

10. Boden, M. *Jean Piaget*. New York: Penguin, 1980, pp. 120–122.

11. *Ibid.*, p. 75.
12. Fraisse, P. *The Psychology of Time*. New York: Harper and Row, 1963.
13. Piaget, *The Child's Conception of Time*, pp. 245–253.
14. Fraisse, *op. cit.*
15. Piaget, Time perception in children, *op. cit.*, p. 212.
16. Cohen, J. *The Psychology of Time in Health and Disease*. Springfield, Ill.: Charles C. Thomas, 1967.
17. Cohen, J., Hamel, E. M., Sylvester, J. D. An experimental study of comparative judgements of time. *British Journal of Psychology* **45**:108–114, 1954.
18. Helson, H., King, S. M. The tau effect: An example of psychological relativity. *Journal of Experimental Psychology* **14**:202–217, 1931.
19. Cohen, J., Ono, A. The hare and the tortoise: A study of the tau effect in walking and running. *Acta Psychologica* **21**:387–393, 1963.
20. DeLong, A. Phenomological space time: Toward an experimental relativity. *Science* **217**:681–683, 1981.

CHAPTER 4

1. Douglas, M. *Purity and Danger*. New York: Praeger, 1969.
2. Evans-Pritchard, E. E. *The Nuer*. Oxford: Clarendon Press, 1940, pp. 104–108.
3. Bohannon, P. Concepts of time among the Tiv of Nigeria. *Southwestern Journal of Anthropology* **9**:251–263, 1953.
4. Beidelman, T. Kagura time reckoning: An aspect of the cosmology of an East African people. *Southwestern Journal of Anthropology* **19**:9–20, 1963.
5. Bohannon, *op. cit.*
6. Leach, E. *Rethinking Anthropology*. Dublin: Athlone Press, 1961, p. 124.
7. Eliade, M. *The Myth of the Eternal Return*. New York: Pantheon, 1954, pp. 89–90.
8. Stanner, W. E. H. *Australian Aboriginal Mythology*. Canberra: Australian Institute of Aboriginal Studies, 1975.
9. Leach, *op. cit.*, p. 126.
10. Kluckhohn, R. R. Deviant and variant value orientation, in *Nature, Society, and Culture*. Edited by C. Kluckhohn, A. Murray, and D. M. Schneider. New York: Knopf, 1965, pp. 342–367.
11. Nakamura, H. Time in Indian and Japanese thought, in *The Voices of Time*. Edited by J. T. Fraser. New York: Braziller, 1966, pp. 77–91.

12. Doob, L. W. *The Patterns of Time*. New Haven, Conn.: Yale University Press, 1971.

13. Levine, R. V., West, L. J., Reis, H. T. Time and punctuality in the U.S. and Brazil. *Journal of Personality and Social Psychology* **38**:541–550, 1980.

14. Gay, J., Cole, M. *The New Mathematics and an Old Culture*. New York: Holt, Rinehart and Winston, 1967.

15. Robbins, M., Kilbride, D. L., Bukenya, J. M. Time estimation and acculturation amongst the Baganda. *Perceptual and Motor Skills* **26**: 1010, 1968. (The author is indebted in this section to Doob, *op. cit.*)

16. Lehmann, H. Time and psychopathology. *Annals of the New York Academy of Sciences* **138**:789–821, 1967.

17. Doob, *op. cit.* p. 104.

18. Knapp, R. H., Garbutt, J. T. Time imagery and the achievement motive. *Journal of Personality* **26**:426–434, 1958.

19. Doob, *op. cit.*, p. 100.

20. Ostheimer, J. M. Measuring achievement motivation among the Chagga of Tanzania. *Journal of Social Psychology* **78**:17–30, 1969.

21. Knapp, R. H. Achievement and aesthetic choice, in *Motives in Fantasy, Action, and Society*. Edited by J. W. Atkinson. New York: Van Nostrand, 1958, pp. 367–372.

22. Bochner, S., David, K. H. Delay of gratification, age, and intelligence in an aboriginal culture. *International Journal of Psychology* **3**:167–174, 1968.

23. *Ibid.*, 173.

CHAPTER 5

1. Sackett, D. L. Bias in analytic research. *Journal of Chronic Disease* **32**:51–68, 1979.

2. Natelson, B. H., Haupt, E. J., et al. Temporal orientation and education. *Archives of Neurology* **36**:444–446, 1979.

3. Pope, H. G. , Lipinski, J. F. Diagnosis in schizophrenia and manic depressive illness: A reassessment of the specificity of "schizophrenic" symptoms in the light of current research. *Archives of General Psychiatry* **35**:811–828, 1978.

4. Fischer, F., quoted in Schilder, P. Psychopathology of time. *Journal of Nervous and Mental Diseases* **83**:530–546, 1936.

5. Lanzkron, J., Wolfson, W. Prognostic value of perceptual distortion of temporal orientation in chronic schizophrenics. *American Journal of Psychiatry* **114**:744–746, 1958.

6. Dahl, M. A singular distortion of temporal orientation. *American Journal of Psychiatry* **115**:146–149, 1958.

7. Crow, T., Stevens, M. Age disorientation in chronic schizophrenia. *British Journal of Psychiatry* **133**:137–143, 1978.

8. Melges, F. T. *Time: The Inner Future*. New York: Wiley-Interscience, 1982.

9. Melges, F. T., Fougerousse, C. Time sense, emotions, and acute mental illness. *Journal of Psychiatric Research* **4**:127–140, 1966.

10. Clausen, J. An evaluation of experimental methods of time judgement. *Journal of Experimental Psychology* **40**:756–761, 1950.

11. Siffre, M. *Beyond Time*. New York: McGraw-Hill, 1964.

12. Clausen, *op. cit.*

13. Bindra, D., Waksberg, H. Methods and terminology in studies of time estimation. *Psychological Bulletin* **53**:155–159, 1956.

14. Doob lays out this debate in detail, see *The Patterns of Time*. New Haven, Conn.: Yale University Press, 1971.

15. Melges, *op. cit.*

16. Mezey, A. G., Knight, E. J. Time sense in hypomanic illness. *Archives of General Psychiatry* **12**:184–186, 1965.

17. Cappon, D., Banks, R. Experiments in time perception. *Canadian Psychiatric Association Journal* **9**:396–410, 1962.

18. Mezey. A. G. Cohen, S. I. The effect of depressive illness on time judgement and time experience. *Journal of Neurology, Neurosurgery, and Psychiatry* **24**:269–270, 1961.

19. Clausen, *op. cit.*

20. Melges and Fougerousse, *op. cit.*

21. Pearl, D., Berg, P. Time perception and conflict arousal in schizophrenia. *Journal of Abnormal and Social Psychology* **66**:332–338, 1963.

22. Wallace, M., Rabin, A. I. Temporal experience. *Psychological Bulletin* **57**:213–236, 1960.

23. Lehmann, H. Time and psychopathology. *Annals of the New York Academy of Sciences* **138**:789–821, 1967.

24. Orme, J. E. Time estimation and the nosology of schizophrenia. *British Journal of Psychiatry* **112**:37–39, 1966.

25. Lehmann, *op. cit.*

26. Melges, F. T., Freeman, A. M. Temporal disorganization and inner–outer confusion in acute mental illness. *American Journal of Psychiatry* **134**:874–877, 1977.

27. Lehmann, *op. cit.*, p. 804.

28. Wallace, M. Future time perspectives in schizophrenia. *Journal of Abnormal and Social Psychology* **52**:240–245, 1956.

29. Dilling, C. A., Rabin, A. I. Temporal experience in depressive states and schizophrenia. *Journal of Consulting Psychology* **31**:604–608, 1967.

30. Schlossberg, A. Time perspective in schizophrenics. *Psychiatric Quarterly* **43**:22–34, 1969.

31. Wallace and Rabin, *op. cit.*

32. Kastenbaum, R. The direction of time perspective. *Journal of General Psychology* **73**:189–201, 1965.

33. Wallace, *op. cit.*, 1956.

34. Geschwind, N. Presentation, American Psychiatric Association Convention, New York, May 1983.

35. Mezey and Cohen, *op.cit.*

36. Wyrick, R., Wyrick, L. Time experience during depression. *Archives of General Psychiatry* **34**:1441–1443, 1977.

37. Mezey and Cohen, *op. cit.*

38. Strauss, E. W. Disorders of personal time in depressive states. *Southern Medical Journal* **40**:254–259, 1947.

39. F. Wooley in Strauss, *Ibid.*

40. Lehmann, *op. cit.*

41. Mezey and Cohen, *op. cit.*

42. Wyrick and Wyrick, *op. cit.*

43. Kirstein, L., Bukberg, J. Temporal disorganization of primary affective disorder. *American Journal of Psychiatry* **136**:1313–1316, 1979.

44. Schilder, *op. cit.*

45. Mezey and Knight, *op. cit.*

46. *Ibid.*

47. Redl, F., Wineman, D. *The Aggressive Child*. Glencoe, Ill.: Free Press, 1957, pp. 119–120.

48. Siegman, A. W.. The relationship between future time perspective, time estimation, and impulse control in a group of young offenders and in a control group. *Journal of Consulting Psychology* **25**:47–75, 1961. *See also*, Effects of auditory stimulation and intelligence on time estimation in delinquents and non-delinquents. *Journal of Consulting Psychology* **30**:320–328, 1966.

49. Brock, T. C., Del Guidice, C. Stealing and temporal orientation. *Journal of Abnormal and Social Psychology* **66**:91–94, 1963.

50. Klineberg, S. L. Changes in outlook on the future between childhood and adolescence. *Journal of Personality and Social Psychology* **7**:185–193, 1967.

51. Stein, K. B., Sabin, R., Kulik, J. A. Future time perspective: Its relation to the socialization process and the delinquent role. *Journal of Consulting Psychology* **32**:257–264, 1968.

52. Getsinger, S. H., Leon, R. Impulsivity, temporal perspective, and posthospital adjustment of neuropsychiatric patients. *Journal of Psychology* **103**:221–225, 1979.

53. Doob, *op. cit.*

54. For a full discussion of this matter, see Doob, *op. cit.*

55. Brown, L. B. Religious belief and judgement of brief duration. *Perceptual and Motor Skills* **20**:33–34, 1965.

56. Campos, L. Relationship between time estimation and retentive personality traits. *Perceptual and Motor Skills* **23**:59–62, 1966.

57. Langer, J., Wapner, S., Werner, H. The effect of danger upon the experience of time. *American Journal of Psychology* **74**:94–97, 1961.

58. Baddeley, A. D. Time estimation at reduced body temperature. *American Journal of Psychology* **79**:475–479, 1966.

59. Fraisse, P. *The Psychology of Time*. New York: Harper and Row, 1963.

60. Friel, C. M., Lhamon, W. T. Gestalt study of time estimation. *Perceptual and Motor Skills* **21**:603–606, 1965.

61. See Ornstein, R. E. *On the Experience of Time*. Baltimore, Md.: Penguin, 1970.

62. Doob, *op. cit.*, p. 170.

63. Cappon and Banks, op. cit.

64. Caine, E. D., Shoulson, I. Psychiatric syndromes in Huntington's disease. *American Journal of Psychiatry* **140**:728–733, 1983.

65. Cottle, T. J., Klineberg, S. L. *The Present of Things Future: Explorations of Time in Human Experience*. New York: Macmillan, 1974.

CHAPTER 6

1. Freud, S. *New Introductory Lectures on Psychoanalysis*. London: Hogarth Press, 1963.

2. Colarusso, C. The development of time sense from birth to object constancy. *International Journal of Psychoanalysis* **60**:243–251, 1979.

3. Bonaparte, M. Time and the unconscious. *International Journal of Psychoanalysis* **21**:427–468, 1940.

4. Fraser, J. T. Temporal levels of reality testing. *International Journal of Psychoanalysis* **62**:3–26, 1981.

5. Bonaparte, *op. cit.*, pp. 465–466.

6. Bonaparte, *ibid.*, p. 438.

7. Bonaparte, *ibid.*, p. 440.

8. Pollack, G. On time, death and immortality. *Psychoanalytic Quarterly* **40**:435–446, 1971.

9. Brenner, C. *Psychoanalytic Technique and Psychic Conflict*. New York: International University Press, 1976.

10. Fraser, *op. cit.*

11. The author is indebted in this section to Colarusso, *op. cit.*

12. Gifford, S. Sleep, time, and the early ego. *Journal of American Psychoanalytic Association* **8**:5–42, 1960.

13. Rapoport, D. *Organization and Pathology of Thought*. New York: Columbia University Press, 1951.

14. Piaget, J. Time perception in children, in *The Voices of Time*. Edited by J. T. Fraser. New York: Braziller, 1966, pp. 202–216.

15. Spitz, R. A. Bridges: On anticipation, duration, and meaning. *Journal of American Psychoanalytic Association* **20**:721–735, 1972.

16. Mahler, M. S. Symbiosis and individuation: The psychological birth of the human infant. *Psychoanalytic Study of the Child* **29**:89–106, 1975.

17. Ames, L. S. The development of the sense of time in the young child. *Journal of Genetic Psychology* **18**:97–125, 1946.

18. Bonaparte, *op. cit.*

19. Fraser, J. T. *The Voices of Time*. New York: Braziller, 1966.

20. Fraser, 1981, *op. cit.*

21. Hartocollis, P. Time and the dream. *Journal of the American Psychoanalytic Association* **28**:861–877, 1980.

22. Loewald, H. W. The experience of time. *Psychoanalytic Study of the Child* **27**:401–410, 1972.

23. Melges, F. T. Time: *The Inner Future*. New York: Wiley-Interscience, 1982.

CHAPTER 7

1. Prigogine, I. Physics and metaphysics. *Advances in Medicine and Biophysics* **16**:241–265, 1977.

2. Borges, J. L. *Labyrinths*. New York: New Directions, 1962.

3. Waddington, C. H. Concluding remarks, in *Evolution and Consciousness*. Edited by E. Jantsch and C. H. Waddington. Reading, Mass.: Addison-Wesley, 1976.

4. Melges, F. T. *Time: The Inner Future*. New York: Wiley-Interscience, 1982.

5. Prigogine, *op. cit.*

Index

ABOUT THE AUTHOR

Matthew J. Edlund, M.D., is Assistant Professor in the Department of Psychiatry and Behavioral Sciences, University of Texas at Houston. He trained in psychiatry at New York University-Bellevue Hospital, and was previously a Clinical Fellow in Medicine at Massachusetts General Hospital. He holds a master's degree in Occupational Health from Harvard.